Minimally Invasive Neurosurgery II

Edited by
B. L. Bauer and D. Hellwig

Acta Neurochirurgica
Supplement 61

Springer-Verlag Wien New York

Professor Bernhard Ludwig Bauer, M.D.
Dieter Hellwig, M.D.
Department of Neurosurgery, Philipps-University Marburg,
Federal Republic of Germany

© 1994 Springer-Verlag/Wien
Softcover reprint of the hardcover 1st edition 1994

Typesetting: Thomson Press, New Delhi, India

Printed on acid-free and chlorine free bleached paper

With 110 partly coloured Figures

ISSN 0065-1419 (Acta Neurochirurgica/Suppl.)

ISBN-13: 978-3-7091-7431-9   e-ISBN-13: 978-3-7091-6908-7
DOI: 10.1007/978-3-7091-6908-7

Preface

Minimally Invasive Neurosurgery is neither a new discipline nor another type of neurosurgery – we have not given birth to a new child. To look back on the history of neuroendoscopy is a humbling but also an illuminating experience. We can see that the trail was clearly blazed by the pioneers Walter E. Dandy[1] and Jason Mixter[2], who performed the first endoscopic plexus extirpation (1922) and third ventriculostomy (1923). 1992 we edited the first update of *Minimally Invasive Neurosurgery (MIN I)*. This supplementum 54 of the Acta Neurochirurgica served to describe the current state of the art in this rapidly developing special field of neurosurgery. First clinical experience using new technologies had been reported. The positive response encouraged us to ask colleagues to present papers on the occasion of a second meeting on endoneurosurgery, held from January 30th – February 1st, 1992, in the Department of Neurosurgery, Philipps University of Marburg, Germany.

The subject matter was limited to the endoscopic anatomy, technical devices and surgical management of disorders suitable for endoscopic procedures. We are grateful to all participants of MIN II. Regretably it was not possible, to obtain papers from all contributors for publication.

As with any multi contributor books, the contributions are not yet balanced. The volume MIN II presents different aspects of that issue. In our opinion the papers are well written and we hope that they are stimulating and informative. It must be also pointed out that the papers presented concerning indication and approaches in different diseases are highly preliminary and long term follow up results are not yet available. The clinical value and the benefit to our patients of this new technical discipline devoted to the use of miniaturized flexible steerable and rigid endoscopes within the skull and spine must be proved against the contemporary well established standards of microsurgery. We entirely assure with the demand of Stephen J. Haines[3] for properly planned and conducted randomized clinical trials, following a track parallel to that used for new drugs. Up to now, most of the contributions are in a pilot stage (phase I). Only a few are in phase II of development with estimations of the rate of success and complications associated with the procedure. The golden standard is a prospective randomized clinical trial (phase III) for every special indication. Unfortunately MIN II is yet not ready to fulfil Dr. Haines' hope that this volume could include a number of well designed clinical trials documenting the clinical effectiviness of neuroendoscopic treatment. The ongoing project is now to develop the design for multicenter randomized clinical trials and to outline the requirements and basic concepts for the use of endoscopes in neurosurgical practice.

Computerized tomography and Magnetic Resonance Imaging has contributed greatly to the diagnosis of intracranial and intraspinal space occupying lesions.

With the gift and with the promise of the tools of modern manoeuverable and multifunctional instruments we can hope to close the gap between versatility and efficiency of neuroendoscopy as compared with modern microsurgical techniques. This revised and updated second volume MIN II is dedicated to critical readers and reviewers of the neurosurgical society.

Marburg, December 1994   Bernhard L. Bauer, Dieter Hellwig

[1] Dandy WE (1922) An operative procedure for hydrocephalus. Bull Johns Hopkins Hosp 33: 189–190

[2] Mixter WJ (1923) Ventriculoscopy and puncture of the floor of the third ventricle. Preliminary report of a case. Boston Med Surg J 188: 277–278

[3] Haines SJ (1994) Neuroendoscopy. Crit Rev Neurosurg 4: 67–72, Springer, Berlin Heidelberg New York Tokyo

Contents

Listed in Current Contents

Acta Neurochir (1994) [Suppl] 61: 1–12

Minimally Invasive Endoscopic Neurosurgery—a Survey

B.L. Bauer and **D. Hellwig**

Department of Neurosurgery, Philipps University Marburg, Federal Republic of Germany

Summary

In 1989 we introduced "endoscopic stereotaxy" as a new operative procedure into neurosurgery. This technique was first scheduled to optimize stereotactic biopsy. In its further development it proved to be effective for other indications. We choose the term "*Minimal Invasive (Endoscopic) Neurosurgery (MIEN)*" for these interventions. Minimal invasive endoscopic techniques are applied preferably for diagnostic and therapeutic interventions on preformed or pathological cavities of the central nervous system. The indications are

— endoscopic-stereotactic biopsy of space-occupying lesions,
— ventriculoscopy and endoscopic ventriculostomy in diagnosis and treatment of hydrocephalus,
— endoscopic evacuation of cystic space occupying lesions,
— endoscopic evacuation of intracerebral haematoma,
— endoscopic evacuation of septated chronic subdural haematoma,
— endoscopic evacuation of subacute or chronic brain abscesses,
— endocavitary syringostomy.

Our results with minimal invasive endoscopic interventions for different indications are encouraging when compared to conventional microsurgical techniques. We have performed more than 300 minimal invasive endoscopic interventions. The mortality rate was below 1 %, the operative morbidity was below 2 %.

Keywords: Minimal invasive neurosurgery; neuroendoscopy; stereotactic neurosurgery; brain tumour; hydrocephalus; brain abscess; intracerebral haematoma; chronic subdural haematoma; syringomyelia.

Introduction

Neuroendoscopy is not new. In the beginning of this century Dandy, Mixter, Putnam, and Scarff[17,55,62,68] performed neuroendoscopic interventions mainly in treatment for hydrocephalus either with choroid plexus coagulation or third ventriculostomy. The main disadvantage at that time was the large diameter of the endoscopes and the lack of suitable surgical instruments.

A renaissance of neuroendoscopy took place after the development of the Hopkins lens system in 1960.

Auer, Griffith, and Guiot[2,20,23] performed neuroendoscopic interventions for treatment of various lesions. In 1986 Griffith summarized neuroendoscopic techniques and termed the field endoneurosurgery[21]. With the development of fiberoptics the field of indications was enlarged again. Fukushima and Ogata were neurosurgical pioneers in the use of flexible, steerable neuroendoscopes[18,58]. Today technological advances in the development of neuroendoscopes and auxillary instruments make it possible to perform complex neuroendoscopic procedures through a minimal invasive burrhole approach without major tissue traumatization. In analogy to the term minimally invasive surgery, coined by Wickham and Fitzpatrick in 1990[75], we have defined these procedures as *Minimally Invasive (Endoscopic) Neurosurgery (MIEN)*. MIEN refers to neurosurgical interventions in which, by use of endoscopes larger openings of intracranial or intraspinal spaces can be avoided[5,31]. It provides an exiting prospect for some aspects of neurosurgical practice, but does not reach the established standards of safety in current microneurosurgery. There is still a large gap, which has to be bridged between future requirements and today's reality of endoneurosurgery, not only to patients and their surgeons but also to those concerned with health policy and ethical issues. The development from macrosurgery to microsurgery is now nearly completed. The next step to MIEN is a current project. The instrumental requirements and the conditions suitable for an endoscopic approach in neurosurgical practice has to be discussed in depth. MIEN must be evaluated according to the standards of classical microsurgical approaches. It offers obvious benefits to patients and to neurosurgeons, but the present situation is characterized not only by a clearcut list of indications but also by unsolved problems of the desirable level of safety and versatility of the instrumentation available compared with the contemporary state of the art in microsurgery.

Neuroendoscopic operations are not small scale, nor are they operations which should be performed by beginners. MIEN requires clinical experience of the whole of microsurgery as a basis for endoscopic operations. This is the only way in which the patient can be spared from wrong indications for surgery and a poor technique, and the only way of enabling definitive control of complications. The position is obscured by various factors such as the ready accessibility of diagnostic methods—computed tomography (CT) and magnetic resonance imaging (MRI)—and the apparently easy performance of the operation. Critical analysis and monitoring (also in training) is therefore urgently necessary. It is astonishing, how many indiscriminate allegations can be found in the literature. The general thoughtlessness in accepting the printed word as an established fact and in this way "collecting evidence" as well as the indiscriminate consideration of case reports from the older literature, disregarding our present knowledge, entail major dangers and lead to overrating of the method. It is therefore understandable that many experienced neurosurgeons are critical about any attempt to introduce endoscopic techniques in neurosurgery

Instrumentation

Endoscopes

At present there are many different flexible, steerable, or rigid endoscopes (encephaloscopes) available for neuroendoscopic interventions. The specifications vary, and one must choose according to the planned operative use. It is useless to discuss whether to employ flexible or rigid endoscopes in special situations; the optimal solution is to have both available. The advantage of the rigid endoscope is the optimized optic quality, whereas the flexible endoscope is a "multipurpose-instrument", with the advantage of steerability and due to this with an enlarged field of possible uses. It has to be mentioned that today flexible endoscopes with diameters of 0.6 to 1.5 mm including working channels are used for spinal neuroendoscopy. Furthermore the first clinical trials with 3D-endoscopes with diameters of 4–6 mm are being performed in our department.

In conclusion we state, that there are specific indications for the application of rigid and/or flexible endoscopes with different diameters. Both systems are complementary in modern neuroendoscopy.

Endoscopy Stabilization and Guiding Devices

Neuroendoscopic interventions require a safe and fixed stabilization and guiding system. At the moment there is a lack of sufficient stabilization and guiding devices. Some neuroendoscopic centers use the "Greenberg retractor" or the "peel-away sheet", which in our opinion have the disadvantage of poor stability. For free-hand

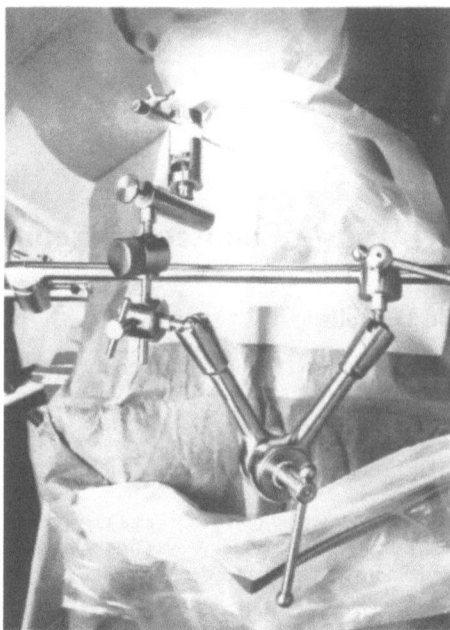

Fig. 1. Marburg Neuroendoscopy Fixation and Guiding device

endoscopy we have developed a special Fixation and Guiding System (*Marburg Neuroendoscopy Fixation and Guiding Device*) (Fig. 1). This system consists of a special self-retaining arm, which provides the necessary stability to the whole endoscopic system during the operative procedure and frees the endoscopist from fatiguing holding work. The endoscopic working depth is regulated by micrometer screws. The metallic guiding system consists of bougies and guiding tubes of different diameters and lengths. The main advantage of our system is, that undesired, uncontrolled movements and manoeuvers during the operation can be avoided, thereby preventing damage to brain tissue. The Fixation and Guiding System is compatible with conventional stereotactic systems.

Supplementary Instruments

Today a variety of micro-instruments for neuroendoscopic interventions are available. However, a special sorted instrumentation set does not exist. In recent years we have worked out, which kind of instruments are required for minimal invasive endoscopic interventions. Some of those are taken from other endoscopic disciplines (gastroenterology) and adapted to neurosurgical demands. Basic instrumentation for minimal invasive endoscopic interventions includes: Microforceps and microscissors for biopsy and dissection of cysts and abscess membranes; grasping forceps to remove cyst material and foreign bodies; balloon catheters for cystostomy or ventriculostomy (Fig. 2). Haemostasis can be performed using monopolar RF, bipolar RF (Fig. 3), or laser energy. Ultrathin bare laser fibers are used for coagulation, vaporization, or cutting of tissues. Intraluminal ultrasound can be used in combination with neuroendoscopical interventions for special cases with intracerebral space-occupying lesions especially in cystic processes. *Dynamic digital subtraction ventriculography* (DDSV) is a useful tool for intraoperative quality control of interventions on the ventricular system. With DDSV we are able to prove restoration of CSF-circulation in

Fig. 2. Balloon catheter for third ventriculostomy

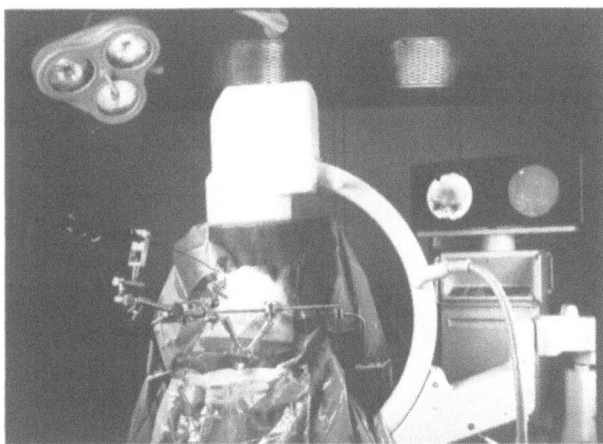

Fig. 4. Stenoscope for intra-operative dynamic digital substraction ventriculography

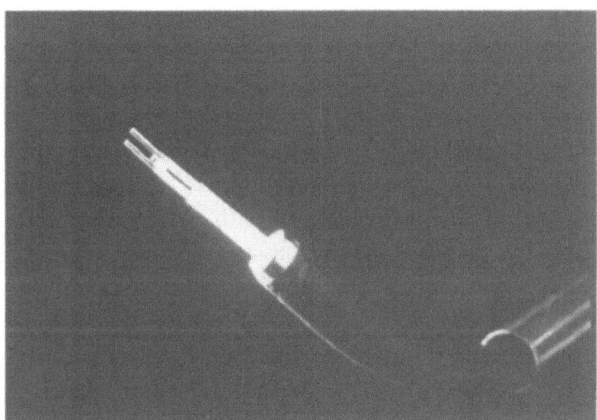

Fig. 3. Bipolar microcoagulation probe

Operating Room Set-up

The complexity of operative technique with video- and laser-technology dynamic, digital picture generating processes demands a highly specialized neuroendoscopic team. Each member plays an important role in the initiation and completion of a successful neuroendoscopic team. The neuroendoscopic team is composed of the neurosurgeon, the anaesthesiologist, instrument nurse, circulating nurse for laser application, and a nurse for camera and video control. The use of the different technologies makes a strict placement and working control of the equipment necessary. Because of limited space in the operating room intra-operative changes of placement are difficult. The surgeon, in the sitting position, has to be reached easily by the instrument nurse, direct visual control of the video screen and the DDSV as well as of the laser unit must be possible. During the intervention direct contact between the anaesthesiologist and the surgeon is indispensable (Fig. 5).

Documentation

It is necessary to have a form of computer-aided documentation for endoscopic interventions. The main goal of such documentation is to obtain an overview about the total number of neuroendoscopic intervention performed, the indications, operative techniques, complications, and results. In addition, data from centers worldwide involved in neuroendoscopy should be available and evaluated.

obstructive hydrocephalus and to control patency of third ventriculostomy (Fig. 4).

At the moment computerized motor-driven supplementary instruments are under development. In future, many additional specialized instruments for minimal invasive endoscopic neurosurgery will be available which will enlarge the spectrum of its usefulness.

Videorecording and Display

Neuroendoscopic operative techniques engender changes of the neurosurgeon's spatial view. In contrast to microsurgical techniques the course of the intervention cannot be controlled directly at the operative site but at the video-screen (the next step will be virtual neurosurgery, where the surgeon imagines to part actively in the interactions between the pathological process in three-dimensional space and the computerized operative tools).

The equipment for intraoperative videorecording and display includes: camera, camera tuner, light source, video-screen, videorecorder. In special cases we use equipment for 3D-endoscopy recording and display.

Indications, Patient Selection, and Results

Today indications for minimal invasive endoscopic interventions are not defined conclusively. This is due to the only brief period since the start in this field which widens or alters operative techniques and strategies. On the other hand, as mentioned before, a generally accepted circumscribed instrumentation set is not available and therefore the lack of useful instruments in some parts limits the extension of neuroendoscopic interventions.

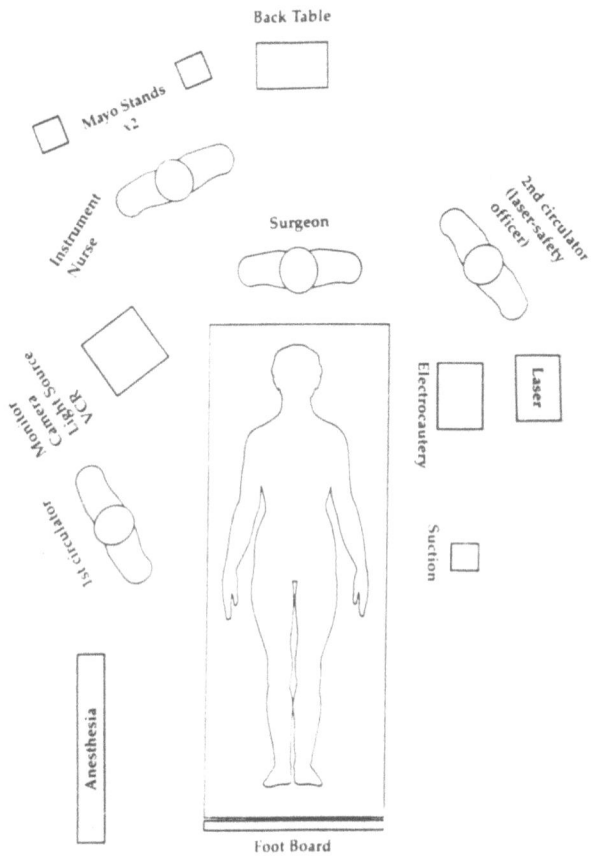

Fig. 5. Operating room set-up

As with all innovative operative techniques the state of testing and quality assurance is a still ongoing project. However, in recent years some indications for minimal invasive endoscopic interventions have been worked out. These are mainly interventions on preformed or pathological cavities of the central nervous system.

We perform minimal invasive neurosurgical procedures for

1. endoscopic treatment of hydrocephalus[5-7],
2. endoscopic stereotaxy[28-30,32],
3. endoscopic evacuation of cystic processes[8,33],
4. endoscopic evacuation of intracerebral haematomas[6,7,31],
5. endoscopic evacuation of septated subdural haematomas[5,6],
6. endoscopic evacuation of brain abscesses[6,7],
7. endoscopic interventions on the spine[5,6].

Endoscopic Treatment of Hydrocephalus

The ventricular system is a preformed intracerebral space and predestinated for neuroendoscopic inter-

ventions. This was first recognized by L'Espinasse, who performed in 1910 the first plexus cauterization[46]. At this time and in the decades afterwards all the trials to treat hydrocephalus with endoscopic interventional techniques failed because there had been a lack of useful instruments. With the development of CSF-diversion systems, treatment of hydrocephalus became standardized.

However, the complications related to shunting systems, such as dysfunction, thrombosis, infection, overdrainage and slit-ventricle syndrome, have lead to a reintroduction of endoscopic operative techniques in the treatment of hydrocephalus, especially of third ventriculostomy.

Third Ventriculostomy

In occlusive hydrocephalus CSF resorption mechanisms are not affected. Consistently it is clear so-called "inner shunting" techniques should be used. In 1923 Mixter performed the first third ventriculostomy in a child with congenital hydrocephalus[55]. He perforated the floor of the third ventricle with a rigid endoscope. The intervention was successful. Beside Mixter[55], Scarff[68] was one of the outstanding advocates of this operative method.

Since then different techniques for third ventriculostomy have been applied, the percutaneous puncturing[10,24,34,67], the stereotatic method[39,43], and endoscopic techniques[15,40,41]. Ventriculostomy with a flexible endoscope combined with a rigid guiding device was first performed by the Marburg group.

Jones, who has contributed to this issue is experienced in the field of endoscopic third ventriculostomy[40,41]. His selection criteria are: patients with aqueductal stenosis or other forms of noncommunicating hydrocephalus; (b) the third ventricle must be of adequate width—at least 7 mm in size, no anatomical contraindications to the procedure such as large massa intermedia or a tiny floor of the third ventricle; (d) previous radiotherapy may present a contra-indication; (e) third ventriculostomy is not indicated in cases of communicating hydrocephalus.

Most authors choose the anterior part of the floor of the third ventricle for fenestration (Fig. 6). A possible alternative in anatomical difficult situation is the fenestration of the lamina terminalis.

Discussion. Today percutaneous third ventriculostomy is obsolete, because of the high fatality rates.

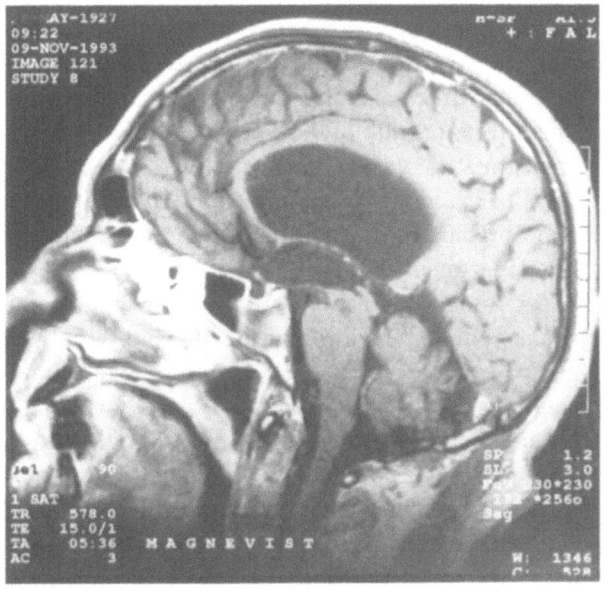 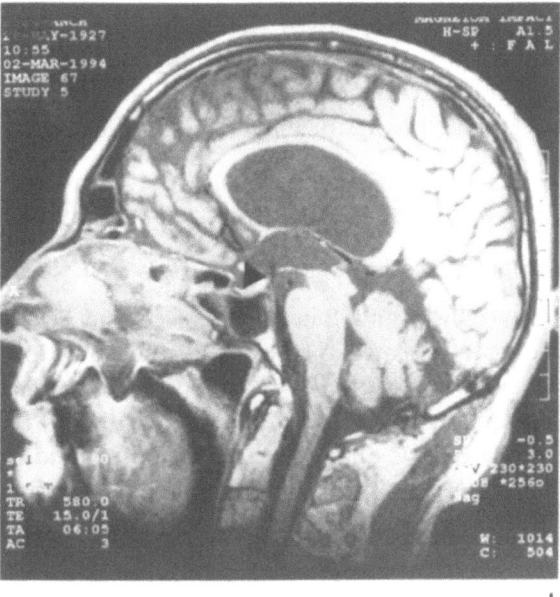

a   b

Fig. 6. (a) Sagittal MRI: Occlusive hydrocephalus caused by benign non-tumorous aqueduct stenosis. (b) MRI control examination: After third ventriculostomy a "flow void" phenomenon at the floor of the third ventricle (black arrow) as a sign of CSF flow into the interpeduncular cistern is visible

Even stereotactic methods have a mortality rate about 5%[43]. Considering our experience with endoscopic third ventriculostomy it has to be remarked, that it is nearly impossible, to reach the proper area for third ventriculostomy at the floor of the third ventricle by "blind" puncture without major traumatization of brain tissue or structures related to the foramen of Monro. Furthermore it has to be stressed that stereotactic technique for third ventriculostomy as described by Kelly[43] can be hazardous. Anatomy of the third ventricle and the interpeduncular fossa in patients with chronic hydrocephalus is variable. The floor of the third ventricle can be elevated by the basilar artery or some of its branches. These partly subtle variants could be missed by pre-operative neuroradiological examination, and "blind" stereotactically guided ventriculostomy could lead to dislocation or vascular injury, which could not be managed using the stereotactic approach. After extensive neuro-anatomical studies of the third ventricle ([63], see Riegel *et al.*, pp. 54–56) we are convinced, that third ventriculostomy should be performed exclusively using endoscopic techniques. Instrumental fenestration of the floor of the third ventricle can be managed by different techniques. Jones[40] performs third ventriculostomy with the rigid endoscope without auxillary instruments. Manwaring[48] and Heilman[27] use the so-called

saline torch, Cohen[15] the laser-probe for fenestration of the third ventricle. Some authors recommend to apply balloon catheters for widening of the artificial porus. We use the flexible steerable endoscope together with the rigid guiding system to perform third ventriculostomy. The floor of the third ventricle is opened with the help of bipolar radiofrequency or "bare" laser probe. After this a Fogarty catheter is introduced to widen the artificial porus. The efficacy of the intervention is controlled intra-operatively by digital dynamic subtraction ventriculography. Postoperative CCT- or MRI-examination should be assessed critically for successful control of third ventriculostomy, because 60% of the patients showed a marked improvement of neurological deficits after third ventriculostomy without change of ventricular size[66].

Compared to others our operative technique for third ventriculostomy using the flexible steerable endoscope has the following advantages:

1. The success of the intervention is independent of anatomical variants. Dislocations can be corrected without major tissue damage.
2. Uncontrolled movements are avoided by application of the endoscopy fixation and guiding system.
3. Application of the microbipolar probe offers a space controlled energy delivery compared to laser

energy. Ventriculostomy by blunt perforation should be avoided.

4. Intra-operative digital dynamic subtraction ventriculography proves patency of third ventriculostomy.

Endoscopic Stereotaxy

In 1989 Hellwig and Bauer introduced endoscopic techniques in stereotatic neurosurgery and defined the combination of these two operative techniques "endoscopic stereotaxy"[28, 29]. At this time there were only single reports about the application of rigid or flexible endoscopes during stereotactic interventions. The idea to introduce endoscopy into stereotactic neurosurgery was born by the endeavour to have a intra-operative visual control during the stereotactic procedure. Endoscopic stereotactic operative techniques are used in different lesions mainly to hit the pathological intracerebral processes unerringly through a minimal invasive burrhole approach. At the moment different endoscopic-stereotactic approaches are performed

1. the precoronal frontal approach,
2. the infratentorial suboccipital approach,
3. the combined frontal-transventricular approach for interventions on intraventricular or ventricle-related processes, where the "region of interest" is reached under stereotactic conditions. The further course of the intervention is done by "free-hand" technique under endoscopic control. The operative techniques of endoscopic stereotaxy are described in detail elsewhere[7, 29, 32]. The indications for stereotactic-endoscopic interventions are various. Cystic intracerebral lesions, brain abscesses and intracerebral haematomas are operated on using this technique.

Endoscopic-Stereotactic Biopsy

A special indication which has to be discussed, are endoscopic stereotactic biopsies of space-occupying processes. The advantages of the combination of endoscopic and stereotactic operative techniques are: (a) early recognition of postbioptic haemorrhages, which allows immediate haemostasis; (b) intra-operative differentiation between normal and pathological brain tissue; and (c) complete evacuation of cystic

intracerebral processes and abscesses under visual control. To date we have performed 146 stereotactic biopsies with an operative morbidity of 2.0% and operative mortality of 0.6%.

Histopathological diagnosis was established in 85% of cases, which is more than 10% higher than that obtained by conventional stereotactic biopsies in our department[53, 54].

Discussion. Operative mortality of stereotactic biopsies is about 1% and operative morbidity between 1 and 6%[22, 57, 59, 73, 74]. With the use of endoscopes we could drop the mortality and morbidity rates to the lower level of the international norm. The main advantage of the endoscopic-stereotactic biopsy technique is the possibility of intra-operative haemostasis, if a postbioptic haemorrhage occurs. Reliability of histopathological diagnosis of the stereotactic biopsies is given between 70 and 94%[22, 73, 74]. We have raised the number of correct histopathological diagnosis in our department from 75% (stereotactic technique) to 85% (endoscopic-stereotactic technique). This is due to the fact, that we are able to differentiate endoscopically between normal and pathological tissue and that it is very easy to take biopsies from cyst walls under visual control which is a tremendous advantage against simple stereotactic biopsy.

Cystic Intracranial Lesions

Cystic intracerebral lesions are a domain or neuroendoscopic interventions. This group of pathological intracerebral cavities include colloid cysts, cystic gliomas, cystic craniopharyngiomas, other dysembryogenetic tumours and space-occupying arachnoid cysts. The aim of the neuroendoscopic intervention is the evacuation of the cystic part of the tumour to decrease raised intracranial pressure (ICP). In many cases these procedures precede microsurgical resection of the solid tumour parts. In colloid cysts or solitary cystic craniopharygiomas, endoscopic interventions have a high success rate without further management. A total of 70 cystic lesions have been operated on in our department by endoscopic, partly in endoscopic-stereotactic technique. In some cases an Omaya reservoir was placed for repeated aspiration of the cyst contents. If the cyst contents are of low viscosity there is no problem aspirating them, whereas in high-viscosity cysts aspiration may fail, and the contents have to be removed with microforceps or microscissors. It is not necessary to remove the whole

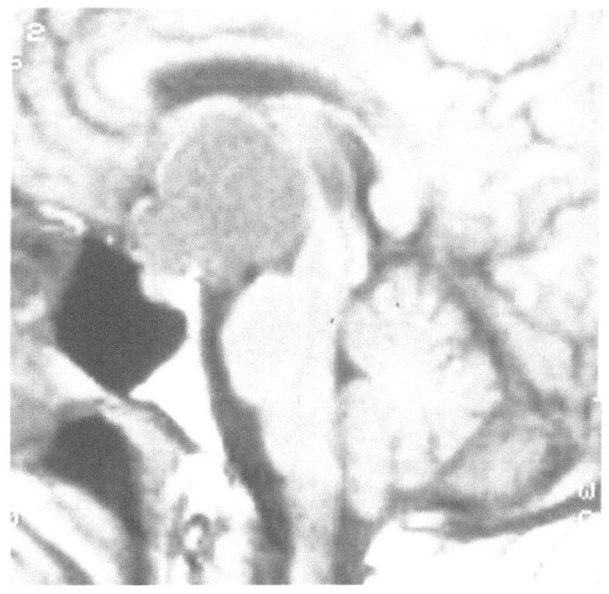 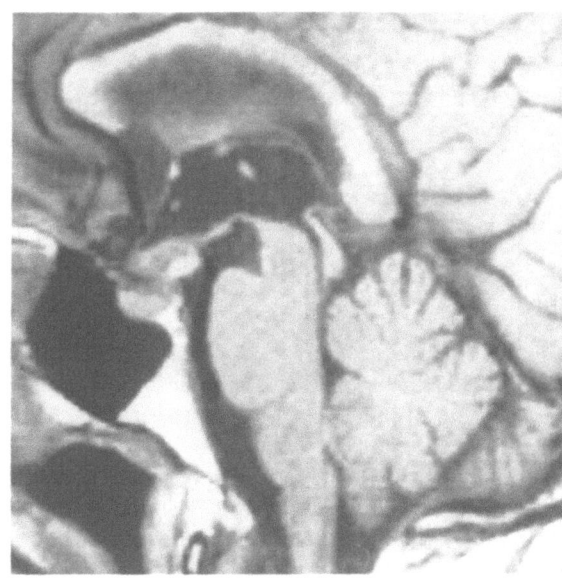

a   b

Fig. 7. (a) Sagittal MRI: cystic astrocytoma of the pituitary stalk, bulging into the third ventricle. (b) MRI control examination: after endoscopic stereotactic cyst evacuation the third ventricle is decompressed

cyst membrane because the possibility of cyst reformation is extremely low. Usually we vaporize the cyst membrane by laser. There was no fatality, operative mortality was 1.4%.

Discussion. Our collection of patients with intracerebral cystic lesions was most heterogenous. The endoscopic intervention on cystic anaplastic gliomas, glioblastomas, or cystic metastases is palliative and performed to reduce raised ICP in combination with the application of reservoir-systems, and to establish a histopathological diagnosis. In benign cystic lesions, especially in intraventricular cysts, the neuroendoscopic intervention is performed curatively.

Cystic lesions of the third ventricle (colloid cyst, pineal cyst, cystic craniopharyngioma) should be discussed in detail. Using microsurgical techniques different operative approaches are necessary to treat these lesions at various locations[12,69,71,72], whereas with the frontal precoronal burr-hole approach we are able to reach all parts of the third ventricle with the flexible steerable endoscope. Operative fatality and morbidity are low, compared to conventional microsurgical techniques. The main discussion point is, that remaining parts of the cyst membrane could lead to a recurrence of the space occupying cyst[47,50]. In our experience and in long term follow-up studies of patients with intraventricular cystic lesions[14,16,60], which where operated on by stereotactic technique,

this assumption could not be verified. Endoscopic stereotactic operative techniques are a minimal invasive option in the treatment of intraventricular space occupying cystic lesions (Fig. 7a, b).

Spontaneous Intracerebral Haematoma

Intracerebral haematoma can be subdivided in two groups—primary hypertensive haemorrhages, and secondary haemorrhages caused by other disease or drug induced haemorrhages.

In 70% intracerebral haematoma are caused by hypertension. Mortality of spontaneous intracerebral haematoma is estimated at 80%. There is still controversy as to whether massive intracerebral haematomas should be operated on or be treated conservatively. The main problem with operative treatment of intracerebral haematoma is, that the intervention is an added trauma to the primary disastrous effect of the bleeding. For this reason in recent years different minimal invasive operative techniques have been proposed. The stereotactic[4,11,35,45] and endoscopic[3] evacuation techniques of intracerebral haematoma have to be emphasized. In 1990 we have combined these two methods to the endoscopic stereotactic method[3,31].

The indications for this intervention are

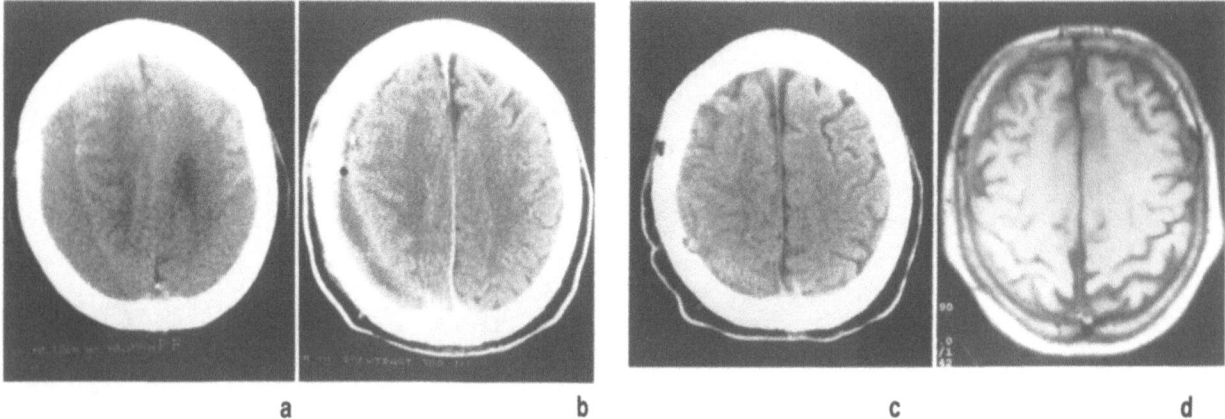

a   b   c   d

Fig. 8. (a) CCT: Septated right hemispheric subdural haematoma. (b) Contrast CCT control examination: septated residual haematoma after burrhole drainage treatment with surrounding capsule. (c, d) CCT and MRI: complete haematoma removal after neuro-endoscopic stereotactic intervention

1. reduction of raised ICP,
2. avoidance of secondary traumatization of brain tissue,
3. avoidance of secondary neurological deterioration,
4. shortening of convalenscence.

Discussion. In 1961 McKissock could show in a randomized study with 180 patients, that immediate evacuation of spontaneous intracerebral haematoma could not lower mortality rate[52]. This result was corroborated by Kanno[42]. Mohadjer *et al.* who summarized mortality and morbidity rates from 13 studies with operative treatment versus conservative treatment of spontaneous intracerebral haemorrhages and found out that there is no evident advantage for operative treatment[56].

Stereotactic aspiration treatment of intracerebral haematoma was first performed by Komai in 1974[45]. From 1978 to 1986 Itakura *et al.* operated on 241 intracerebral haematoma stereotactically. They concluded: the operative procedure should be performed between six hours and three days after onset of the bleeding[38]. The use of urokinase as fibrinolytic agent during stereotactic haematoma evacuation was a further attempt to improve operative results[38,45]. Operative fatality (10,7 %) and morbidity (10,4 %) of this technique are very low compared to conventional surgical treatment. Endoscopic techniques were introduced by Auer[3]. He used rigid endoscopes under ultrasonic guidance. In his study postoperative long-term result is influenced by the haematoma volume, the age and the pre-operative level of consciousness of the patient. We have adopted Auer's minimal invasive

approach and combined it with stereotactic operative techniques. Our results allow the following conclusions: (a) the intervention should be performed in the first 48 hours after the onset of the bleeding; (b) haematoma volumes of more than 50 ml adversely affect prognosis; (c) more than 50 % of the haematoma volume can be removed using this technique; (d) comatose patients do not benefit from the intervention; (e) in atypical haemorrhages stereotactic biopsy under endoscopic control enables one to clarify the diagnosis. The comparison of different treatment strategies for spontaneous intracerebral haematoma is very difficult, because there are varying opinions about the timing of the operation, the haematoma volume which has to be removed, and the neurological state, which demands immediate intervention.

A randomized multicenter study about the indications for minimal invasive techniques and comparison with conventional operative procedures as well as with conservative treatment is urgently needed.

Septated Chronic Subdural Haematoma

Successful treatment of nonseptated chronic subdural haematoma (CSH) can be easily achieved by burrhole trepanation and the insertion of a subdural silicon catheter. Treatment of septated CSH using this method may be unsuccessful since the subdivision of the haematoma cavity by neomembranes can limit the complete efflux of haematoma fluid. We have operated on 14 patients with septated CSH using endoscopic techniques. Eleven showed complete haematoma removal on the postoperative CCT control examination whereas three patients had to be operated on a second time endoscopically to remove the residual haematoma fluid. There has been one postoperative infectious complication (Fig. 8a–d).

Discussion. Conventional operative treatment of chronic subdural haematoma has been primary craniotomy with consecutive haematoma evacuation. This regime was replaced by twist-drill craniostomy" combined with a closed drainage-system[19,25,49,64,65]. Operative mortality of this method is between 0 and 9.5 %. Some authors report a re-operation rate of up to 20 % using this technique[25].

The indications for re-operation with an extended craniotomy are: Frequent recurrence of the haematoma; space occupying residual haematoma; clotting of the haematoma, space-occupying hemispherical oedema due to residual haematoma. The main problems with these cases are the neomembranes, which divide the haematoma cavity into compartments and hinder the efflux of the haematoma fluid.

In 1991 we introduced neuroendoscopic techniques in operative treatment of septated chronic subdural haematoma in cases where primary "twist-drill craniostomy" has failed. Through a burr-hole approach the haematoma is reached with the flexible steerable endoscope and the neomembranes are cut by microscissors under direct visual control. With this technique we can achieve communication between the haematoma compartments and effect a free efflux through a drainage system. As a consequence of our results we recommend endoscopic operative techniques for evacuation of septated chronic subdural haematoma where primary burr-hole evacuation methods have failed.

Brain Abscess

Brain abscesses are a further group of pathological intracerebral cavities, in which the performance of minimal invasive endoscopic neurosurgical techniques are very useful.

Different methods of treatment of brain abscess have been proposed in recent decades. CT stereotactic evacuation of brain abscess seems to have advantages against conventional operative techniques. Today

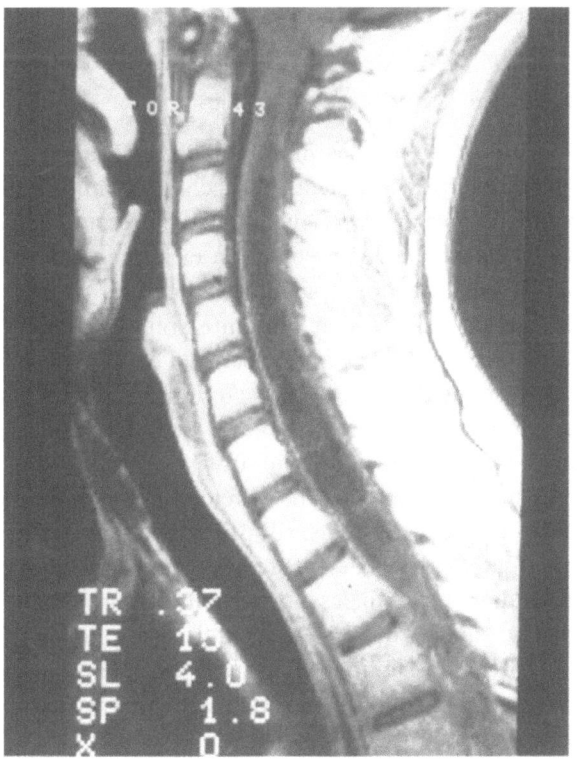
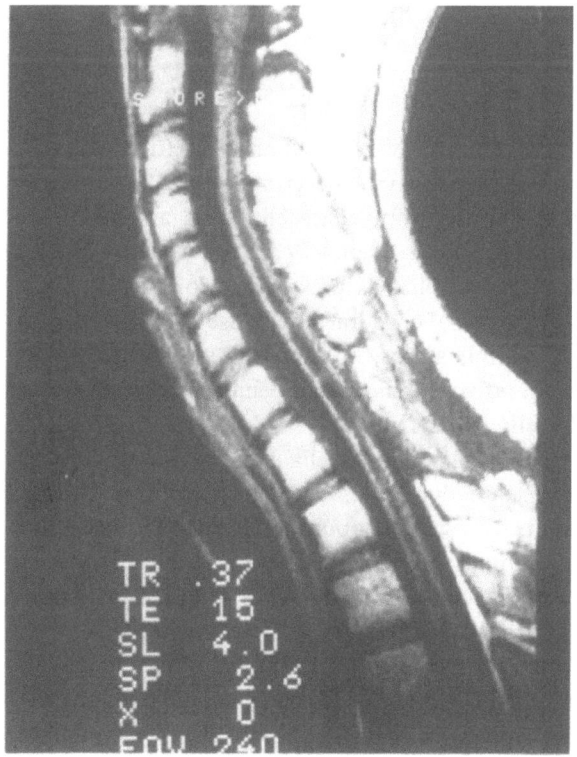

a   b

Fig. 9. (a) Septated syringomyelia of the cervical and upper thoracal spine. (b) After endoscopic intervention only a small remaining cavity is visible

some studies about successful stereotactic treatment of brain abscess have been published[26,70]. The most striking was that of Hasdemir and Ebeling in 1993. They had a mortality rate of 4% after stereotactic aspiration. Postoperative follow-up between three and twelve months showed that 75% of the patients got back to normal life. Only 20% of the patients showed handicaps in managing their daily life.

King and Turney were the first to report encephaloscopic intervention in a brain abscess[44]. We have also used the endoscopic approach and combined it with stereotactic guidance. The operative aim is to reduce acute ICP elevation, sample infectious material for microbiological examination, and attempt to cure the lesion by serial puncturing and aspiration. Indications, operative technique and the results are described in detail on other pages of this issue.

Intraspinal Endoscopy

Intraspinal endoscopy is not as common as intracranial endoscopy. In 1931 Burman was the first who performed endoscopic examinations of the spinal cord[13]. Since then only some reports of spinal endoscopy have been published[9,61].

Hydromyelia, Syringomyelia

In 1989 we introduced together with Hüwel the technique of "endocavitary syringostomy" into spinal endoscopy. Hydromyelia, defined as an increase of fluid in the dilated central canal of the spinal cord, and syringomyelia, as presence of longitudinal cavities in the spinal cord lined by dense gliogenous tissue, are ideal indications for neuroendoscopic techniques. The operative technique had been described by Hüwel and Hellwig in MIN I[31,36].

Discussion. Syringomyelia is an extraordinary complex disease due to its various aetiologies. Aschoff gave an overview of about 1152 publications with a total number of 3077 cases. At present 15 different operative techniques with 40 variants are performed. The overall results of operative treatment show a slight advantage of the shunting methods over decompressive techniques[1]. Hüwel *et al.* proposed neuroendoscopic interventions in all cases of septated syringomyelia[37].

In 1993 Hüwel *et al.* published their experiences with 39 patients, who were treated by "endocavitary syringostomy". There were 19 patients with idiopathic syringomyelia, 7 with tumour associated syringomyelia, 4 with tethered cord and 9 patients with posttraumatic syringomyelia. The postoperative outcome was excellent overall. No complications were reported[37]. We have performed neuroendoscopic interventions in six patients with septated syringomyelia. In none of these patients have we seen postoperative impairment of neurological symptoms in the first few months after operation; however, at the moment there is no long-term follow-up.

In our opinion "endocavitary syringostomy" should be reserved for patients, who suffer from idiopathic septated syringomyelia, whereas in all other cases differential diagnostic aspects and differential therapeutic options should be discussed.

Endoscopic Discectomy

Percutaneous endoscopic laser discectomy is increasingly important due to technical advances and the trend to MIEN, with reduced surgical trauma, shorter hospital stay, reduced costs, and early re-integration to work and normal life. This is particularly true with the percutaneous endoscopic treatment of lumbar disc disease as introduced by Mayer and Brock in 1989[51].

Conclusion

This survey about "Minimal Invasive (Endoscopic) Neurosurgery", its indications, techniques and results intends to give an overview about the present state of the art in this field. Based on an experience of more than 300 neuroendoscopic interventions for treatment of different intracranial and intraspinal lesions the results are compared and discussed with regard to conventional neurosurgical techniques. However, at the moment there are so many activities and tendencies concerning minimal invasive techniques in neurosurgery, especially the development of computer guided operative systems, roboting, and virtual medicine, that in the near future this summary may be out of date. Nevertheless, Minimally Invasive Neurosurgery I and II must be seen as a description of the recent real possibilities in endoscopic neurosurgery and we are glad that our idea, which was born in 1989, is still an ongoing project.

References

1. Aschoff A, Kunze S (1993) 100 years syrinx-surgery—a review. Compiled by E. Donauer. Acta Neurochir (Wien) 123: 157–159
2. Auer LM, Holzer P, Heppner R (1988) Endoscopic neurosurgery. Acta Neurochir (Wien) 90: 1–14
3. Auer LM, Deinsberger W, Niederkorn K, et al (1989) Endoscopic surgery versus medical treatment for spontaneous intracerebral hematoma: a randomized study. J Neurosurg 70: 530–535
4. Backlund EO (1971) A new instrument for stereotactic brain tumour biopsy. Acta Chir Scand 137: 825–827
5. Bauer BL, Hellwig D (1992) Minimally invasive neurosurgery I. Acta Neurochir (Wien) [Suppl] 54
6. Bauer BL, Hellwig D (1994) Current Endoneurosurgery. In: Bauer BL et al (eds) Advances in neurosurgery, Vol 22. Springer, Berlin Heidelberg New York Tokyo, pp 113–120
7. Bauer BL, Hellwig D (1994) Intracerebral and Intraspinal Endoscopy. In: Schmideck HH, Sweet WH (eds) Operative neurosurgical techniques, 3rd Ed. Saunders, Philadelphia, in press
8. Bauer BL, Hellwig D, Sweet WH, Schmideck HH (1994) The management of intracranial arachnoid, suprasellar and rathke's cleft cysts. In: Schmideck HH, Sweet WH (eds) Operative neurosurgical techniques, 3rd Ed. Saunders, Philadelphia, in press
9. Blomberg R (1985) A method for epiduroscopy and spinaloscopy. Presentation of preliminary results. Anesth Analg 66: 177–180
10. Broggi G, Franzine A, Peluchetti D, et al (1974) Transcallosal third ventriculochiasmatic cisternostomy: a new approach to hydrocephalus. Surg Neurol 2: 109–114
11. Broseta J, Gonzalez-Darder J, Barcia-Salorio JL (1982) Stereotactic evacuation of intracerebral hematomas. Appl Neurophysiol 45: 443–498
12. Bruce JN, Stein BM (1992) Infratentorial approach to pinealis tumors. In: Wilson CB et al (eds) Neurosurgical procedures. Personal approaches to classic operations. Williams and Wilkins, Philadelphia, pp 63–75
13. Burman MS (1931) Myeloscopy or the direct visualization of the spinal cord. J Bone Joint Surg 13: 695
14. Caemaert J, Abdullah J (1993) Diagnostic and therapeutic stereotactic cerebral endoscopy. Acta Neurochir (Wien) 124: 11–13
15. Cohen AR (1993) Endoscopic laser third ventriculostomy. N Engl J Med 328: 552
16. Cohen AR (1993) Endoscopic ventricular surgery. Pediatr Neurosurg 19: 127–134
17. Dandy WE (1992) Cerebral ventriculoscopy. Bull John Hopkins Hospital 33: 189
18. Fukushima T, Ishijima, Hirakama, et al (1973) Ventriculofiberscope: a new technique for endoscopic diagnosis and operation. J Neurosurg 38: 251–256
19. Gilsbach J, Eggert HR, Harders A (1980) Externe geschlossene Drainagebehandlung des chronischen Subduralhämatoms nach Bohrlochtrepanation. Unfallchirurgie 3: 183–186
20. Griffith HB (1975) Technique of fontanelle and persutural ventriculoscopy and endoscopic ventricular surgery in infants. Childs Brain 1: 359–363
21. Griffith HB (1986) Endoneurosurgery: endoscopic intracranial surgery. In: Symon L et al (eds) Advances and technical standards in neurosurgery, Vol 14. Springer, Wien New York, pp 2–24
22. Grunert P, Ungersböck K, Bohl J, et al (1994) Results of 200 intracranial stereotactic biopsies. Neurosurg Rev 17: 59–66
23. Guiot G, Rougerie J, Fourestier M, et al (1963) Une nouvelle technique endoscopique: explorations endoscopiques intracraniennes. Presse Med: 1225–1228
24. Guiot G (1973) Ventriculocisternostomy for stenosis of the aequeduct of Silvius. Acta Neurochir (Wien) 28: 274–289
25. Harders A, Eggert HR, Weigel K (1982) Behandlung des chronischen Subduralhämatomes mit externer geschlossener Drainage. Neurochirurgia 25: 147–152
26. Hasdemir MG, Ebeling U (1993) CT-guided stereotactic aspiration and treatment of brain abscesses. An experience with 24 cases. Acta Neurochir (Wien) 125: 58–63
27. Heilman CB, Cohen AR (1991) Endoscopic ventricular fenestration using a "saline torch". J Neurosurg 74: 224–229
28. Hellwig D, Eggers F, Bauer BL (1990) Endoscopic stereotaxis. Preliminary results. Stereotact Funct Neurosurg 54, 55: 418
29. Hellwig D, Bauer BL (1991) Endoscopic procedures in stereotactic neurosurgery. Acta Neurochir (Wien) [Suppl] 52: 30–32
30. Hellwig D, Bauer BL, List-Hellwig E, et al (1991) Stereotactic endoscopic procedures on processes of the cranial midline. Acta Neurochir [Suppl] 53: 23–32
31. Hellwig D, Bauer BL (1992) Minimally invasive neurosurgery by means of ultrathin endoscopes. Acta Neurochir (Wien) [Suppl] 54: 63–68
32. Hellwig D, Bauer BL (1993) Diagnostik und Therapie intrakranieller Tumoren. Endoskopische Stereotaxie ein neues Operationsverfahren. TW Neurologie Psychiatrie 7: 347–356
33. Hellwig D, Riegel T, Bauer BL, et al (1994) Endoneurosurgery of skull base processes. The frontal transventricular approach. In: Samii M (ed) Skull base surgery. Karger, Basel, in press
34. Hoffmann HJ, Harwood-Nash D, Gilday DL (1980) Percutaneous third ventriculostomy in the management of noncommunicating hydrocephalus. Neurosurgery 7: 313–321
35. Hokana M, Tanizaki K, Masruo K, et al (1993) Indications and limitations for CT-guided stereotaxic surgery of hypertensive intracerebral haemorrhage, based on the analysis of postoperative complications and poor ability of daily living in 158 cases. Acta Neurochir (Wien) 125: 27–33
36. Hüwel NM, Perneczky A, Urban V (1992) Neuroendoscopic technique for operative treatment of septated syringomyelia. In: Bauer BL, Hellwig D (eds) MIN I. Acta Neurochir (Wien) [Suppl] 54: 59–62
37. Hüwel NM, Perneczky A, Urban V (1993) Neuro-endoscopic techniques in operative treatment of syringomyelia. Acta Neurochir (Wien) 123: 216
38. Itakura T, Komai N, Nakai E, et al (1988) Stereotactic evacuation of hypertensive intracerebral hematoma using plasminogen activator. Surgical technique and long-term results. In: Suzuki J (ed) Advances in surgery for cerebral stroke. Springer, Berlin Heidelberg New York Tokyo, pp 443–448
39. Jack CR Jr, Kelly PJ (1989) Stereotactic third ventriculostomy: assessment of patency with MR imagery. AJNR 10: 515–522
40. Jones RFC, Stening WA, Brydon M (1990) Endoscopic third ventriculostomy. Neurosurgery 26: 86–92
41. Jones RFC, Teo C, Stening WA, et al (1992) Neuroendoscopic third ventriculostomy. In: Crone KR, Manwaring KH (eds) Neuroendoscopy, Vol 1. Liebert, New York, pp 91–96
42. Kanno T, Sanno H, Shinomiya Y, et al (1984) Role of surgery in hypertensive intracerebral hematoma. J Neurosurg 61: 1091–1099
43. Kelly PJ (1991) Stereotactic third ventriculostomy in patients with nontumoral adolescent-adult onset of aqueductal stenosis and symptomatic hydrocephalus. J Neurosurg 75: 865–873

44. King JEJ, Turney F (1936) Brain abscess. Evolution of the methods of treatment. Ann Surg 103: 647

45. Komai W, Doi E, Moriwaki, et al (1986) Stereotactic evacuation of hypertensive thalamic hematoma using plasminogen activator (urokinase). Neurol Surg 14: 249–256

46. L'Espinasse VL (1992) In: Davis L (ed) Hydrocephalus and spina bifida. Principles of Neurological Surgery. Lea and Febinger, Philadelphia, pp 438–447

47. Lewis AI, Crone K, Taha J, et al (1994) Surgical resection of third ventricle colloid cysts. Preliminary results comparing transcallosal microsurgery with endoscopy. J Neurosurg 81: 174–178

48. Manwaring KH (1992) Endoscopic Ventricular fenestration. In: Crone KR, Manwaring KH (eds) Neuroendoscopy, Vol 1. Liebert, New York, pp 79–89

49. Markwalder T (1981) Chronic subdural hematoma: a review. J Neurosurg 54: 637–645

50. Mathiessen T, Grane P, Lindquist C, et al (1993) High recurrence rate following aspiration of colloid cysts in the third ventricle. J Neurosurg 78: 748–752

51. Mayer HM, Brock M (1989) Percutaneous lumbar discectomy. Springer, Berlin Heidelberg New York Tokyo

52. McKissock W, Richardson A, Taylor GM, et al (1961) Primary intracerebral haemorrhage. A controlled trial of surgical and conservative treatment in 180 unselected cases. Lancet 11: 221–226

53. Mennel HD, Roßberg C, Lorenz H, et al (1989) Reliability of simple cytological methods in brain tumour biopsy diangosis, Neurochirurgia 32: 129–132

54. Mennel HD, Hellwig D, Bauer BL (1994) Ergebnisse und Zuverlässigkeit stereotaktisch und endoskopisch gewonnener Probebiopsate bei Hirntumoren. Zentralbl Neurochir 55: 79–90

55. Mixter WJ (1923) Ventriculoscopy and puncture of the third ventricle. Boston Med Suurg J 188: 277–278

56. Mohadjer M (1993) Computed tomographic—stereotactic evacuation and fibrinolysis of hypertensive intracranial hematoma. Fibrinolysis 2: 43–48

57. Mundinger F, Birg W (1984) Stereotactic biopsy fo intracranial processes. Acta Neurochir (Wien) [Suppl] 33: 219–224

58. Ogata M, Ishikawa T, Horide R, et al (1974) Encephaloscope: basic study. J Neurosurg 22: 288–291

59. Ostertag CH B (1988) Reliability of stereotactic brain tumour biopsy. In: Lunsford LD (ed) Modern stereotactic neurosurgery. Nijhoff, Boston, pp 129–136

60. Ostertag CH B (1990) Surgical techniques in the management of colloid cysts of the third ventricle—the stereotactic endoscopic approach. In: Symon L et al (eds) Advances and technical standards in neurosurgery, Vol 17. Springer, Wien New York, pp 143–149

61. Pool JL (1942) Myeloscopy: intrathecal endoscopy. Surgery 11: 169–182

62. Putnam T (1934) Treatment of hydrocephalus by coagulation of the choroid plexus: description of a new instrment and preliminary report of results. N Engl J Med 22: 1373–1376

63. Riegel T, Hellwig D, Bauer BL, et al (1994) Endoscopic anatomy of the third ventricle. In: Bauer BL, et al (eds) Advances in neurosurgery, Vol 22. Springer, Berlin Heidelberg New York Tokyo, pp 121–125

64. Robinson RG (1984) Chronic subdural hematoma: surgical management in 133 patients. J Neurosurg 61: 263–268

65. Rosenbluth PRB, Arias B, Quartetti EV (1962) Current management of subdural hematoma. JAMA 179: 115–118

66. Saint-Rose C (1992) Third ventriculostomy. In: Crone KR, Manwaring KH (eds) Neuroendoscopy, Vol 1. Liebert, New York, pp 47–62

67. Sayers NP, Kosnik EJ (1976) Percutaneous third ventriculostomy. Experience and technique. Childs Brain 2: 24–30

68. Scarff JE (1970) The treatment of nonobstructive (communicating) hydrocephalus by endoscopic cauterization of the choroid plexus. J Neurosurg 33: 1–18

69. Shucart WA, Stein BM (1978) Transcallosal approach to the anterior ventricle system. Neurosurgery 3: 339–343

70. Stapleton SR, Bell BA, Uttly D (1993) Stereotactic aspiration of brain abscesses. Is this the treatment of choice? Acta Neurochir (Wien) 121: 15–19

71. Stein BM (1971) The infratentorial supracerebellar approach to pineal lesions. J Neurosurg 35: 197–202

72. Symon L, Pell M (1990) Surgical techniques in management of colloid cysts of the third ventricle. The transcortical approach. In: Symon L et al (eds) Advances and technical standards in neurosurgery, Vol 17. Springer, Wien New York, pp 122–130

73. Thomas DGT, Nouby RM (1989) Experiences in 300 cases of CT-directed stereotactic surgery for lesion biopsy and aspiration of hematoma. Br J Neurosurg 3: 321–326

74. Voges J, Schröder R, Trerues H, et al (1993) CT-guided and computer assisted stereotactic biopsy. Technique, results, indications. Acta Neurochir (Wien) 125: 142–149

75. Wickham J, Fitzpatrick J (1990) Minimally invasive surgery. Br J Surg 77: 721

Correspondence: B.L. Bauer, M.D., D. Hellwig, M.D., Department of Neurosurgery, Philipps University Marburg, D-35033 Marburg, Federal Republic of Germany.

Acta Neurochir (1994) [Suppl] 61: 13–19

Technology Assessment of Endoscopic Surgery

E. Neugebauer[1], **B.M. Ure**[2], **R. Lefering**[1], **E.P. Eypasch**[2], and **H. Troidl**[2]

[1] Biochemical and Experimental Division and [2] Surgical Clinic, II. Department of Surgery, University of Cologne, Cologne, Federal Republic of Germany

Summary

Endoscopic surgery is considered a milestone in the evolution of surgical technique in nearly all fields of surgery. However, the inappropriate use of the new technology in medicine has also been heavily criticised. Systematic technology assessment of endoscopic surgical techniques is mandatory to prove the real benefits and complications, so defining the indications for their appropriate use. This article describes methods of technology assessment suitable for endoscopic techniques with emphasis on relevant endpoints for surgeons and patients.

The general stages of a comprehensive technology assessment include:

1. feasibility (safety and technical performance)
2. efficacy (patient benefits in pioneering places)
3. effectiveness (patient benefits in average hospitals in the community as a whole) and
4. economic evaluation (cost-benefit analyses).

We used the example of laparoscopic cholecystectomy to describe the methods of technology assessment. A cohort study on 500 patients revealed that laparoscopic cholecystectomy is as safe as the conventional standard open technique. The results on efficacy strongly support the hypothesis of more comfort and less trauma with the endoscopic technique. Major endpoints evaluated were postoperative pain, convalescence, fatigue and quality of life. Data on effectiveness and ecomonics are still in a "premature" state and should be the subject of further analyses.

It is concluded, that other disciplines such as neurosurgery should evaluate their endoscopic surgical techniques according to the rules of technology assessment outlined in this paper.

Keywords: Technology assessment; laparoscopic cholecystectomy; endoscopic surgery.

Introduction

In the field of general surgery, endoscopic surgery is considered a milestone in the evolution of surgical techniques. Laparoscopic cholecystectomy and appendectomy are on the way to becoming standard operations in many surgical clinics all over the world, and like a bushfire new endoscopic operative techniques are introduced in all fields of surgery. Numerous reports indicate that endoscopic surgery offers significant advantages over the corresponding conventional operation[2,12,15,28,32,35,38]. It has been postulated that due to the minimal trauma endoscopic surgery would improve discomfort and in particular pain as compared to the conventional operation[23,32,36].

However, most reports on laparoscopic techniques are restricted to technical performance and safety alone[29,34]. Jennett[18] pointed out that "it is one thing to demontrate that immediate technical objectives can be achieved, another to show that their achievements lead to an improvement in outcome for patients". Consequently, systematic technology assessment of endoscopic surgical techniques is mandatory to prove the benefit for the patient. The aim of this article is to describe methods of technology assessment suitable for endoscopic techniques. We intend to point out the relevant endpoints for surgeons *and* patients and to demonstrate the current state of technology assessment for laparoscopic cholecystectomy, the most widespread endoscopic technique.

General Stages of Technology Assessment

In general, of comprehensive technology assessment includes four steps that have been described by Jennett[18] (Table 1). Unfortunately there are contradictory definitions of these terms in the literature. The definitions given in Table 1 are promoted by the Committee for Evaluating Medical Technologies in Clinical Use published in 1985[25].

The first stage after development of a new technique is to establish that short-limited goals are reachable. The initial evaluation is commonly referred to as *feasibility* which means assessment of safety and technical performance.

Table 1. *General stages in Technology Assessment (Jennett 1986[18])*

Feasibility	safety, technical performance
Efficacy	patient benefit in pioneering places
Effectiveness	in average hospital in community patient benefit as a whole
Economical appraisal	cost-benefit comparison

impact on — medical practice / resource allocation / society

Table 2. *Pros and Cons for a Randomized Controlled Trial as a Method for Evaluating Feasibility and Efficacy of Laparoscopic Cholecystectomy*

Pro
- Prompt and reliable evaluation of safety and efficacy necessary
- No reliable comparison without concurrent controls
- Positive/negative results due to selection bias
- Untested use increases uncertainty about its value
- Enthusiasm and technical success intensifies the illusion of better treatment
- Postponement of trial initiation may eliminate a definitive therapeutic trial (Chalmers)

Contra
- Proper information on feasibility (safety, technical performance) is lacking
- Surgical technique is not standardized
- Constant development of surgical skill and surgical instruments within a short time (learning curve)
- No data on relevant endpoints (comfort of the patient) available
- Techniques to assess relevant study endpoints not proven

The next stage, testing *efficacy*, deals with the application of a technology already found to be feasible. Patients' outcome can no longer be evaded but is a "sine qua non" for this stage of assessment. This must include both, immediate and late mortality, the reduction of specific symptoms, the quality of the patient's life or other innovative endpoints such as pain and fatigue. Up to this point the new technology and the related patient benefits were exclusively assessed in pioneering centers.

These early results lead to the next step, the assessment of the general applicability of the technique. *Effectiveness* means assessing the technology in average hospitals in the community as a whole. This means that effectiveness refers to what happens when there is widespread application. This step is absolutely necessary before a technology can be recommended as a standard procedure.

It is difficult to imagine a technology that would be justified whatever it costs. Therefore, the final step before adoption of a new technology is *economic evaluation*. It deals with both, inputs and outputs, costs and the consequences of the activities. In general, the costlier the intervention is, the greater its demonstrated effectiveness must be before it is considered an acceptable alternative, at least by care givers.

Methods of Technology Assessment in Endoscopic Surgery

We have a substantial body of methods available for assessment of medical technologies[25] and the question is which techniques are most appropriate for the evaluation in different stages of development. Mosteller[25] stated that the most rigorous way to determine the usefulness of a new technology involves randomly allocating subjects to either receive or not to receive the technology and then comparing their eventual outcomes. The randomized controlled trial has become the accepted gold standard for demonstrating

the true therapeutic effects in evaluating most forms of treatment and one proposal has even been to "randomize the first patient"[5,6].

A randomized controlled trial on laparoscopic versus conventional cholecystectomy would reveal proper and reliable information on safety and efficacy without selection bias. There are, however, some strong arguments for not running a randomized controlled trial in the evolutionary phase of a new operation, especially when a new technique requires training. In the phase of introduction of laparoscopic cholecystectomy there was no proper information concerning feasability and the operative technique was not standardized. In addition the surgical skill and the instruments underwent continuous development. There were no data available on relevant endpoints dealing with the comfort of the patient and finally the techniques to assess these endpoints were not proven in patients with symptomatic gallbladder disease.

One alternative for the evaluation of a technique in the evolutionary as well as in a later phase is a prospective observational study resulting in a computerized data-base. This offers potentional additional research opportunities. One advantage over randomized controlled trials is the fact that big data-bases give the opportunity to detect rare events. They are a good basis for quality control and they offer assessment of feasibility and efficacy in the real world allowing economic calculations as well.

However, the use of data-bases represent difficult methodological problems. Data often are inaccurate or incomplete. There are significant management problems especially when a data-base is built up in more than one center. Rough estimates are easy to obtain but they will often not be very helpful to answer specific questions. One inherent problem of databases is that the diagnostic procedures, the treatment modalities and the criteria for evaluation may change with time and differ between centers. Further problems may arise with data protection and the methods used to analyze the data. They are much more open to evaluating bias than hypothesis which are set up before running a randomized controlled trial.

The decision on the methods to be used for the assessment of a medical technology depend on the stage of evaluation. Randomized controlled trials or observational studies are suitable for assessing feasibility or efficacy. Effectiveness is evaluated preferably by other methods such as quantitative evaluation of different surgeons in one hospital, literature reviews, meta-analyses, group-judgement methods, surveillance studies and systematic follow-up[25].

For the economic evaluation several methods have been proposed[11,39,40]. *Cost-evaluation* as the easiest method only estimates the costs of a technique and therefore represents a partial form of economic appraisal. When the consequences of alternatives are assumed to be equal and only the costs are compared, the evaluation is called a *cost-minimization analysis*. *Cost-effectiveness* is the analysis of costs which are related to a single common effect, e.g. life years gained per dollar spent. In this analysis no attempt is made to value the consequences. In the *cost-utility analysis* the consequences of alternatives are measured in time units adjusted by health utility weights. In general terms this means that one can assess for example the quality of life-years gained, not just the crude number of years. The *cost-benefit analysis* represents the broadest form of economic evaluation. Attempts are made to ascertain different beneficial consequences of a technique in monetary values. It should be stressed that not all methods of economic evaluation are suitable for every technique and that clear distinctions often are not possible.

The example: Technology Assessment of Laparoscopic Cholecystectomy

The motivation to start with endoscopic surgery was the hypothesis that this technology leads to less trauma and consequently more comfort for the patient.

Table 3. *Major Endpoints and Results of the Prospective Observational Study of the First 100 Patients with Laparoscopic Cholecystectomy*[27]

Mortality		0	Pain intensity (VAS scale)
Complications			1. postop. day 33
intra-op.		1	2. postop. day 23
– postop., major		2	3. postop. day 10
– postop., minor		12	discharge 7
Time of operation (min)			Nausea (n)
Patient 1–50		120	day of operation 51
Patient 51–100		90	1. postop. day 24
Days in hospital (mean,			2. postop. day 8
range)			3. postop. day 1
– pre-operative	2	(1–9)	Vomiting (n)
– postoperative	3	(1–15)	day of operation 22
			1. postop. day 6
			2./3. postop. day 0

This was the case for the introduction of minimal invasive surgery in numerous fields such as orthopaedics, gynaecology, general surgery, or neurosurgery. Consequently the assessment of efficacy is of greater importance than feasability as the latter alone is not sufficient to convince surgeons and patients for using this technology. In this kind of situation feasibility and efficacy have to be evaluated in parallel as soon as possible. We were confronted with this problem after the first laparoscopic cholecystectomy had been performed in October 1989 in our clinic.

Phase I—feasibility and efficacy in patient 1 to 100: Weighing the pros and cons listed in Table 2 it became apperent to us that the randomized controlled trial was not feasable at the early stage of development. Consequently we decided first to initiate a careful prospective observational study to assess feasibility and efficacy in 100 patients (Table 3). This observational study was intended to be replaced by a randomized controlled trial after demonstrating safety and feasibility in 100 patients. We were aware that this group of initial patients was highly selected as most patients had a severity of symptoms of grade zero or one according to the classification of McSherry[22] which means that symptoms were not very severe. Our patients were not recruited exclusively from the Cologne area but from all over Europe.

The major endpoints of this first study period were mortality, time of operation and technical problems, time of hospitalization, intensity of pain, postoperative symptoms as nausea, vomiting, and loss of appetite. There were no deaths and only 3 major complications which were considered avoidable. There were strong benefits with regard to all endpoints concerned with discomfort or disability such as pain, other post-operative complaints, and convalescence[27].

Phase II—feasibility and efficacy in patient 101 to 500: Reassessing the possibility of a randomized controlled trial led to numerous contra-arguments. The analysis of the endpoints estimating efficacy were strongly in favour for laparoscopic cholecystectomy. There was a unanimous agreement that laparosopic cholecystectomy was as safe as the conventional technique in our hands. The therapeutic uncertainty was no longer valid for the surgeons and also the patients were strongly in favour for the new technique. Patients refused to give us informed consent for a randomized trial. In summary we had to accept that ethical restraints led us to consider alternatives to a randomized controlled clinical trial.

Continued surveillance should be maintained at a level which would allow long-term evaluation of both, the old and the new technique. A prospective database seemed to be again a versatile and appropriate alternative. Consequently the most appropriate way to assess feasibility and efficacy of laparoscopic cholecystectomy was to go on with the prospective observational study but with a more extended and standardized protocol. This protocol included 617 variables for every patient documented. Feasibility was still evaluated with 130 variables/patient and a new classification of complications, counting events with negative outcome for the surgeon as well as the patient, was established[35,37]. The endpoints and measures for the assessment of efficacy which included about 400 variables/patient are presented in Table 4.

According to our hypothesis we considered components of the comfort of the patient as described in the next paragraph as the most important endpoints.

Acute postoperative pain was considered the most relevant endpoint. Assessment of pain included the intensity, quality and localization of pain. We exclusively used validated instruments such as the Visual Analogue Scale and the Verbal Rating Scale[17,19] for measuring the intensity and a validated German version of the Short form of the McGill Pain Questionnaire[24,337] for measuring the quality of pain. We carefully documented the use of analgesics. Nausea, vomiting and appetite as parameters of discomfort were recorded at different stages after the operation by a standardized protocol. In addition we assessed which therapeutic modalities such as drainages, tubes, i.v. lines, or thrombosis prophylaxis bothered the patients most. Another important endpoint being considered was fatigue, measured with a Visual Analogue Scale from zero to ten[7]. Quality of life was assessed systematically using several validated instruments. These included a Visual Analogue Scale for life satisfaction and the Gastrointestinal Quality of Life Index[13] which is a questionnaire with 36 items asking for symptoms, emotions, physical and social behaviour.

Statistical analysis of all variables of feasibility and efficacy as well as a group judgment (cholecystectomy study group) was performed after 100 new patients. The results of phase I were confirmed continuously and consequently phase II was continued as planned up to patient 500.

Table 4. *Endpoints and Measures for the Assessment of Efficacy in Phase II (patient 101–500) for Laparoscopic Cholecystectomy*

Criterion	Measures
Pain	
– intensity	VAS, VRS[19]
– quality	Short McGill[22]
– localisation	Questionnaire
Use of drugs	
Pain and distress	Questionnaire[42]
Nausea, vomiting, appetite	Questionnaire
Bothering therapeutic modalities	Questionnaire
Fatigue	VAS (Scale 1–10)[7]
Time of convalescence	
Hospital stay	
Body image	VAS
Quality of life	GI-quality of life index[13]
Bradburn affect balance scale	Self survey
Life satisfaction	VAS

VAS visual analogue scale, *VRS* verbal rating scale.

Table 5. *Incidents of Laparoscopic Cholecystectomy According to Our Classification for the First 500 Patients*[37]

I. *Incident free surgery* No surgical/technical problems and no negative outcome for the patient	54.4 %
II. *Inconsequential-incident surgery* One or more surgical/technical problems, but no negative outcome for the patient (e.g. loss of gallstones, bleeding)	25.5 %
III. *Consequential non-incident surgery* No surgical/technical problems, but one or more negative outcome for the patient (e.g. infection, haematoma)	13.3 %
IV. *Consequential-incident surgery* One or more surgical-technical problems with corresponding negative outcome for the patient (e.g. change of operation proc.)	6.4 %
V. *Death—worst-incident surgery* Lethality includes death in any relation to the operation	0.4 %

Table 6. *Assessment of the Efficacy of Laparoscopic Cholecystectomy: Summary and Main Results After 500 Patients*

- Nausea 21 %, vomiting 14 %, appetite 66 % at 1st postop. day
- Pain intensity less than VAS 50 at 1st postop. day, rapidly decreasing, nearly no pain at 3rd postop. day
- Pain location most signif. at the sites of trocars
- Convalescence period short with rapid return to full activity
- Bothering by therapeutic modalities very low (i.v. lines, drains, tubes, thrombosis prophylaxis)
- Hospital discharge: median 3rd postop. day
- Cosmetics satisfactory
- Fatigue: not changed after 2 weeks
- Quality of life: after 2 weeks signif. better than preop.

Table 7. *Comparison of the Effectiveness of Laparoscopic Cholecystectomy Between Europe and the USA*

Criterion	Europe[8]	USA[34]
LC completed	1191	1446
conversion to open op.	45 (3.6 %)	72 (4.7 %)
technical difficulties	33	28
Complications total	20 (1.6 %)	82 (5.4 %)
serious (with relap.)	9	13
Common bile duct damage	4 (0.33 %)	7 (0.5 %)
Duration of operation (min)	50 (30–90)	90 (19–345)
Death	0	1
Hospital stay	3 (1d–27d)	1.2 (6h–3d)
Treated as outpatient	—	14.2 %
Time to full activity	11 (7–42)	—

The results of the analysis of incidents according to our classification system are presented in Table 5. *Incident-free surgery*, that means no surgical/technical problem and no negative outcome for the patient was present in 54 % of cases. We had *inconsequential-incident surgery* which means one or more surgical/technical problems, but no negative outcome for the patient (e.g. gallstone-loss or bleeding) in an additional 25 % of the patients. *Consequential non-incident surgery* with no surgical/technical problems but a negative outcome for the patient (e.g. infections or haematoma) was present in 13 % of patients. More serious events were classified as *consequential-incident surgery* representing one or more surgical/technical problems with corresponding negative outcomes for the patient (e.g. conversion to conventional operation) and were present in 6.4 %. After 500 operations *death-worst-incident surgery* occured in two patients (0.4 %).

The main results of efficacy were more striking and are presented in Table 6. The mean intensity of pain was less than 50 points on the 100 point scale anytime after the operation and rapidly decreased to nearly zero on the third day. Normal appetite was experienced by 66 % of patients on the first postoperative day and the median hospital discharge was after three days. Patients were generally satisfied with the cosmetic result on follow-up examination two weeks after the operation. In addition, quality of life and fatigue by this time were significantly better than preoperatively. These favourable results lead to our conclusion that laparoscopic compared to conventional cholecystectomy offers numerous advantages and should be promoted.

Phase III—effectiveness: The new laparoscopic technology rapidly spreads throughout the world. The first analyses of the effectiveness of laparoscopic cho-

lecystectomy were included in two publications[8,34] (Table 7). Cushieri reported the main results of seven European centers and the Southern Surgeons Club analyzed 20 centers in the USA, 10 academic and 10 private hospitals. About 80 surgeons were included in both reports, 1200 patients in Europe and 1500 in the USA, half of them in academic and half in private hospitals. The median age was similar in both reports and most patients were operated on for symptomatic gallstone disease, only a few for acute cholecystitis. A conversion to the open operation was about the same percentage. The reported complications showed a difference but this might be due to definitions. There was no difference in common bile duct damage and there was only one death. The hospital stay was short and 14 % of patients were treated as outpatients in the USA. In general, the results were similar in the USA and Europe demonstrating the effectiveness of laparoscopic cholecystectomy for symptomatic gallstone disease. Further analyses on laparoscopic cholecystectomy in average hospitals, however, are mandatory and should be performed for reasons of quality control.

Phase IV—economic evaluation: Economic evaluation is an integrated part of technology assessment and will gain further importance in the future. The economic evaluation of laparoscopic cholecystectomy is still in an early state. Peters *et al.*[29] performed a cost analysis for 93 patients admitted the day of surgery for elective laparoscopic cholecystectomy. The mean charges were 3.620 $ in comparison to a mean of 4.252 $ for previous 58 open cholecystectomies by the same surgeons in the same hospital. This means that the laparoscopic technique saved 600 $ per patient. Cushieri *et al.*[8] reported for the Dundee population a cost saving per patient of 900 pounds which was mainly due to a reduced hospital stay. But these cost analyses are only the first step in economic evaluation. In particular a cost-effectiveness analysis is still mandatory in order to evaluate the benefit of laparoscopic cholecystectomy in terms of saved costs and preferences for the individual patient.

Life Cycle of a New Technology—Where are We Now?

New technologies force changes throughout the system; old procedures are discarded, new ones replace them. Today, endoscopic surgery is considered a fascinating and promising technology and it is on its way to replace conventional technique. In the history of surgery there are numerous examples of new techniques going through a typical 'product life cycle curve'.

The impact of technology assessment on the life cycle of technologies is demonstrated by the example of internal mammary artery ligation as a treatment for angina pectoris[3]. This technique was rapidly introduced in Italy and the United States in the 1950s, with the rationale that the operation would reduce pain by shunting blood into the coronary circulation. Early reports indicated considerable success and thousands of patients underwent this operation. Cobb *et al.*[9] applied the operation to 17 patients; for 8 the real operation was carried out, and these patients reported a 34 percent improvement. Nine patients had a sham operation and these patients reported a 42 percent subjective improvement. By a simultaneous trial Diamond *et al.*[10] achieved similar results. These results led to a rapid discreditation of the technique.

McKinlay[21] describes the complex reaction of doctors, hospitals and administrators to pressures from various quarters when a new technology emerges. It usually starts with *promising reports* of small numbers, rarely failures are mentioned. This has been the case for laparoscopic cholecystectomy in 1987 to 1989[12,28]. Press and media often magnify the significance and soon there is *professional adoption* by some groups or special units for reasons of scientific interest or only prestige. Laparoscopic cholecystectomy was implemented from 1989 and 1990 in numerous centers all over the world and first reports on larger series have been published[4,34].

The next stage is *public acceptance* which leads to the demand that the technology in question should be more widely available. Today, laparoscopic cholecystectomy is on its way to be generally accepted by both, patients and surgeons. But it has to be stressed that good data up to now are limited to studies on safety and the immediate effects on selected patients in special centers. Nevertheless, laparoscopic cholecystectomy has become the *standard procedure* for symptomatic gallstone disease in many centers as well as community hospitals.

At this stage controlled trials often reveal the first doubts that the technique in question is not the panacea as its promotors state. Consequently *professional denunciation* arises from a confrontation between trialists and the product champions. Up to now laparoscopic cholecystectomy has not experienced this stage as all published reports confirm the advantages over the conventional technique. *Discreditation* would be the final step and lead to replacement by the next innovation.

Of course not all technologies go through each stage of such a life cycle. Most frequently they stabilize at the level lower than their peak uptake as limitations of definite benefits become obvious. Therefore, prompt and valid assessment of a medical technology at any stage is most important.

Conclusions

As soon as a new technology is introduced, a systematic and rigorous evaluation is necessary to find out the merits and limitations of the technique in question. This is a scientific and ethical task. Unfortunately, a randomized controlled clinical trial of endoscopic versus conventional techniques often is not possible. This was the case for arthroscopic techniques as well as for laparoscopic cholecystectomy. All results of technology assessment of the first 500 patients that were derived from a cohort study in our clinic revealed that laparoscopic cholecystectomy is as safe as the conventional operation. The results on efficacy strongly support the hypothesis of more comfort and less trauma for the patient. However, it has to be stated that data on effectiveness and economic evaluations still are in too early a state and should be the subject of further analyses. Other disciplines such as Neurosurgery are asked to evaluate their techniques of endoscopic surgery[1,16,26,31] according to the rules of technology assessment. The most relevant endpoint is not the feasibility of the method but the benefit for the patient.

References

1. Auer LM, Holzer P, Ascher PW, Heppner F (1988) Endoscopic neurosurgery. Acta Neurochir (Wien) 90: 1–14
2. Bailey RW, Zucker KA, Flowers JL, Scovill WA, Graham SM, Imbembo AL (1991) Laparoscopic cholecystectomy. Experience with 375 consecutive patients. Ann Surg 213: 531–539
3. Barsamian EM (1977) The rise and fall of internal mammary artery ligation in the treatment of angina pectoris and the lessons learned. In: Bunker JP, Barnes B, Mosteller F (eds) Costs, risks and benefits of surgery. Oxford University Press, New York, pp 212–220
4. Berci G (1991) The Los Angeles experience with laparoscopic cholecystectomy. Am J Surg 161: 382–384
5. Chalmers TC (1975) Randomizing the first patient. Med Clin North Am 59: 1035–1038
6. Chalmers TC (1981) The clinical trial. Milbank Mem Fund 59: 324–339
7. Christensen T, Bendix T, Kehlet H (1982) Fatigue and cardio-respiratory function following abdominal surgery. Br J Surg 69: 417–419
8. Cushieri A, Dubois F, Mouiel F, Mouret P, Becker HD, Buess G, Trede M, Troidl H (1991) The European experience with laparoscopic cholecystectomy. Am J Surg 161: 385–390
9. Cobb LA, Thomas GI, Dillard DH (1959) An evaluation of internal-mammary artery ligation by a double-blind technique. N Engl J Med 260: 1115–1119
10. Diamond EG, Kittle CF, Crockett JE (1958) Evaluation of internal mammary artery ligation and sham procedure in angina pectoris. Circulation 18: 712–716
11. Drummond MF, Stoddard GL, Torrance GW (1987) Methods for the economic evaluation of health care programmes. Oxford University Press, Oxford
12. Dubois F, Ikard PF, Berthelot G, Levard H (1990) Coelioscopic cholecystectomy. Preliminary report of 36 cases. Ann Surg 211: 60–2
13. Eypasch E, Troidl H, Wood-Dauphinee S, Williams JI, Reinecke K, Ure BM, Neugebauer E (1990) Quality of life and gastrointestinal surgery—a clinimetric approach to developing an instrument for its measurement. Theor Surg 5: 3–10
14. Grace PA, Quereshi A, Coleman J, Keane R, McEntee G, Broe P, Osborne H, Bouchier-Hayes D (1991) Reduced postoperative hospitalization after laparoscopic cholecystectomy. Br J Surg 78: 160–162
15. Graves HA, Ballinger JF, Anderson WJ (1991) Appraisal of laparoscopic cholecystectomy. Ann Surg 213: 655–662
16. Hellwig D, Bauer BL (1991) Endoscopic procedures in stereotactic surgery. Acta Neurochir (Wien) [Suppl] 52: 30–32
17. Huskisson EC (1974) Measurement of pain. Lancet ii: 1127–1131
18. Jennett B (1986) High technology medicine. Benefits and burdens. Oxford University Press, Oxford
19. Jensen MP, Karoly P, Braver S (1986) The measurement of clinical pain intensity: comparison of six methods. Pain 27: 117–126
20. McDonald CJ, Hui SL (1991) The analysis of homogous databases: problems and promises. Stat Med 10: 511–518
21. McKinlay JB (1981) From 'promising report' to 'standard procedure': seven stages in the career of a medical innovation. Milbank Mem Fund 59: 374–411
22. McSherry CK, Ferstenberg H, Calhoun F, Lahman E, Virshup M (1985) The natural history of diagnosed gallstone disease in symptomatic and asymptomatic patients. Ann Surg 202: 59–63
23. Meador JH, Nowzaradan Y, Matzelle W (1991) Laparoscopic cholecystectomy: report of 82 cases. South Med J 84: 186–189
24. Melzack R (1987) The short-form McGill pain questionnaire. Pain 30: 191–197
25. Mosteller F (1985) Assessing medical technologies. National Academy Press, Washington
26. Mundinger F, Birg W (1984) Stereotactic biopsy of interacranial processes. Acta Neurochir (Wien) [Suppl] 33: 219–224
27. Neugebauer E, Troidl H, Spangenberger W, Dietrich A, Lefering R, The Cholecystectomy Study Group (1991) Conventional versus laparoscopic cholecystectomy and the randomized controlled trial. Br J Surg 78: 150–154
28. Perissat J, Collet D, Belliard R (1990) Gallstones: laparoscopic treatment—cholecystectomy, cholecystostomy and lithotripsy. Surg Endosc 4: 1–5
29. Peters JH, Ellison EC, Innes JT, Liss JL, Nichols KE, Lomano JM, Roby SR, Front ME, Carey LC (1991) Safety and efficacy of laparoscopic cholecystectomy. Ann Surg 213: 3–12
30. Peters JH, Gibbons GD, Innes JT, Nichols KE, Front ME, Roby SR, Ellison EC (1991) Complications of laparoscopic cholecystectomy. Surgery 110: 769–777
31. Powers SK (1986) Fenestration of intraventricular cysts using a flexible steerable endoscope and the argon laser. Neurosurgery 18: 637–641
32. Schirmer BD, Edge SB, Dix J, Hyser MJ, Hanks JB, Jones RS (1991) Laparoscopic cholecystectomy. Ann Surg 213: 665–676
33. Stein C, Mendl G (1988) The German counterpart to McGill pain questionnaire. Pain 32: 251–255
34. The Southern Surgeons Club (1991) A prospective analysis of 1518 laparoscopic cholecystectomies. N Engl J Med 324: 1073–1078
35. Troidl H, Spangenberger W, Dietrich A, Neugebauer E (1991) Laparoskopische Cholezystektomie. Erste Erfahrungen und Ergebnisse bei 300 Operationen: eine prospektive Beobachtungsstudie. Chirurg 62: 257–265
36. Troidl H, Spangenberger W, Langen R, Al-Jaziri A, Eypasch E, Neugebauer E, Dietrich J (1992) Laparoscopic cholecystectomy: technical performance, safety, and patient benefits. Endoscopy 24: 252–261
37. Troidl H, Eypasch E, Al-Jaziri A, Spangenberger W, Dietrich A (1991) Laparoscopic cholecystectomy in view of medical technology assessment. Dig Surg 8: 108–113
38. Vitale GC, Collet D, Larson GM, Cheadle WG, Miller FB, Perissat J (1991) Interruption of professional and home activity after laparoscopic cholecystectomy among French and American patients. Am J Surg 161: 396–398
39. Warner KE, Hutton RC (1980) Cost-benefit and cost-effectiveness analysis in health care: growth and composition of the literature. Med Care 18: 1069–1084
40. Weinstein MC (1981) Economic assessment of medical practices and technologies. Med Decis Making 1: 309–330
41. Weinstein MC, Coley CM, Richter JM (1990) Medical management of gallstones: a cost-effectiveness analysis. J Gen Intern Med 5: 277–284
42. Zung W (1983) A self-rating pain and distress scale. Psychosomatics 24: 889–893

Correspondence: E. Neugebauer, Ph.D., Biochemical and Experimental Division, II. Department of Surgery University of Cologne, Ostmerheimerstrasse 200, D-51109 Cologne, Federal Republic of Germany.

Acta Neurochir (1994) [Suppl] 61: 20–33

Fundamentals of Laser Science

K.R. Goebel

Surgical Laser Technologies, Darmstadt, Federal Republic of Germany

Summary

The importance of laser application in minimal invasive neurosurgery is emphasized. Physical fundamentals of laser energy are described, and the physics of laser tissue interactions are discussed. Different surgical laser types as the carbon dioxide laser, the argon laser, and the Nd:YAG laser as well as the application mode (contact versus non-contact mode) are compared according application. The main indication for the use of laser in minimal invasive neurosurgery is cutting, coagulation, vaporization, and interstitial irradiation.

Keywords: Laser; neuroendoscopy.

Introduction

Minimal invasive neurosurgical procedures depend on the availability of useful instruments. With the development of ultra-thin laser fibers, which can be used through tiny working channels, the possibility of laser application in neuroendoscopy is possible now. However it is absolutely neccessary that the neurosurgeon who uses laser energy has knowledge about laser physics and the indications and limits of its application.

Laser is an acrynom for Light Amplification by the Stimulated Emission of Radiation. Albert Einstein first postulated the "Quantum Theory of Radiation" in 1917. In this theory he hypothesized the ability to stimulate emission of radiation. Theodor Maiman was the first scientist to build a laser in 1961 while working for Hughes Laboratories (44 years after Einstein first envisioned such a machine).

1. Introduction to Laser Physics

To understand laser, one must first be familiar with the basic principles of light. Laser is a form of light. Light is made up of electromagnetic energy. Electromagnetic energy occupies the "Electromagnetic Spectrum". The electromagnetic spectrum measures frequency and wavelengths given off as energy (radiation) by atomic systems or processes. A small amount of energy is in the visible light spectrum (380 nm–700 nm). The remainder of the radiation lies in the infrared and ultraviolet portion, not visible by the human eye (Fig. 1).

Electromagnetic energy is released as a photon, the basic component of light (including laser light). Photons travel in wave-like patterns (Fig. 2).

Production of Laser Energy

Laser energy is formed by energizing an active lasing medium. The active medium may be in the form of a solid, gas, or liquid. Atoms of the lasing medium occur normally in a resting state. The electrons are in lower orbits around the nucleus (Fig. 3).

With the introduction of energy from an outside source, the atom absorbs the energy causing the electron to rise to a higher or excited state. This external power source can be in the form of optical (flashlamp), another laser, and electrical discharge (Fig. 4).

The electron remains in the excited state for a very brief period before the atom seeks to regain equilibrium by returning to ground state. When the electron falls in orbit, the energy initially absorbed is given off as a photon. The photon or energy bundle has unique properties characteristic to that particular medium. The previous atomic process is termed "Spontaneous Emission of Radiation."

As the photon travels through the medium it strikes another atom in the excited state. The photon is absorbed and causes an increased decay to ground state. During the electron's return to equilibrium two identical photons are emitted. One, the initial photon absorbed, two, that produced by the decaying atom. The

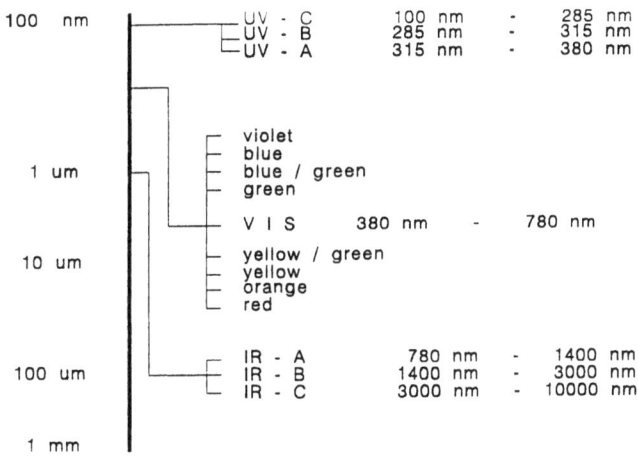

Fig. 1. Electromagnetic spectrum

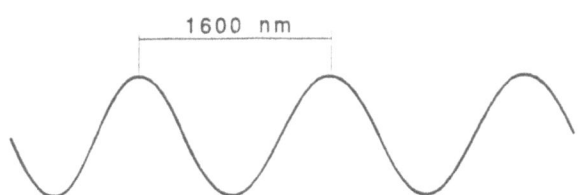

Fig. 2. Distance between the amplitude of 2 consecutive waves, i.e. YAG laser = 1060 nm

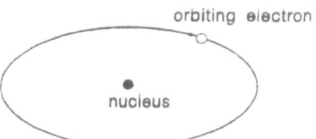

Fig. 3. Resting or "ground state" of the atom

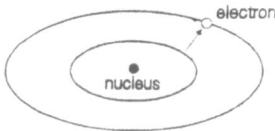

Fig. 4. Electron jumps to higher orbit after absorption of external energy

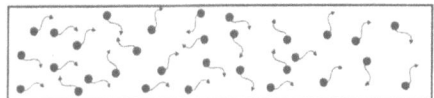

Fig. 5. Spontaneous Emission

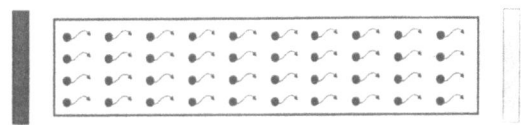

Fig. 6. Stimulated Emission

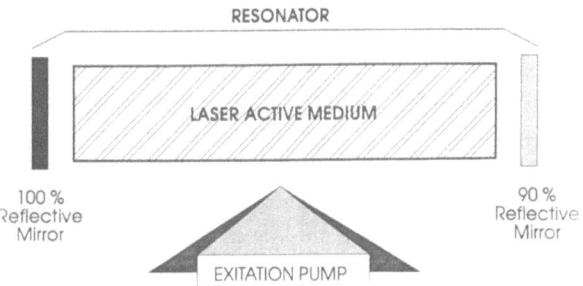

Fig. 7

previous atomic process is termed "Stimulated Emission of Radiation." If a lasing medium contains atoms of molecules that release photons only by spontaneous emission of radiation, the light would be random and evenly distributed in all directions (Fig. 5).

Stimulated Emission takes the process one step further by increasing or amplifying the number of photons on a massive scale. Photons created by stimulated emission establish a pattern in the resonator by reflecting back and forth across the axis of two aligned mirrors. *The intensity of intracavity energy is amplified by reflections between the parallel mirrors.* In order for Stimulated Emission to produce amplification of light, the number of excited atoms must outnumber the atoms or molecules in ground state. This condition is termed "Population Inversion" (Fig. 6).

The laser is comprised of three basic components:

1) the resonator (laser chamber and optical cavity),
2) the active lasing medium,
3) the excitation pump (Fig. 7).

On either end of the laser chamber are mirrors. One mirror is 100 % reflective, the other is 90 % reflective. Parallel to the medium is the external pumping system (i.e. flash lamp). The laser chamber contains the active medium. Each laser is named after the medium used in medical lasers and they are CO_2 (gas), and NdYAG (solid crystal).

Characteristics of Laser Light

The physical properties of the lasing medium and the configuration of the optical chamber are responsible for varying laser wavelengths. Photons given off by stimulated emission have four common properties:

1) monochromatic—the photons travel in the same wavelength or colour,
2) collimation—the wavelengths are spaced equally apart or parallel,
3) coherent—the wavelength are in the same time and space.

Note: Ordinary sunlight or incandescent light has none of the above qualities. It is made up of many random wavelength.

Beam Configuration

As the laser energy exits the cavity or resonator, the beam distributes radially in several configurations. The configuration denotes what type of thermal effect will be produced on tissue. This process is called TEM of Transverse Electromagnetic Mode.

Ideally the energy would be dispersed evenly on tissue for a uniform thermal effect (Fig. 8).

The above is ideal and not easily achieved. The basic intensity profile or configuration of the beam is the Gaussian distribution or TEM_{00} (Fig. 9).

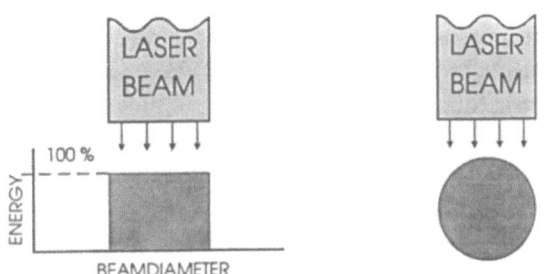

Fig. 8. Left: Beam cross section. Right: Pattern of effect on tissue or impact imprint on tissue

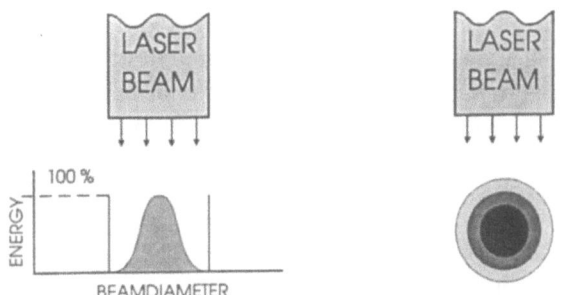

Fig. 9. In this profile the maximum intensity of the beam is in the center and tapers in thermal effect at the periphery

A third mode, the complicated mode, is made up of a variety of intensities on the tissue. This is not as important as the Gaussian mode. The Gaussian mode will be more functional because it allows the smallest focal diameter to be obtained by the particular optical system.

Pulsed and Continuous Wavelengths

Radiant energy delivered by the laser can be in the form of "continuous wave" (cw) or multiple pulses referred to as "pulsed" wave. These two modes of delivery are a function of time.

Continuous wave (cw) constant power levels delivered that are above 0.25 seconds. For example, the SLT CL 60 NdYAG-laser may be delivered in preset pulses from 0.1 second to 10 seconds or the operator may deliver one continuous pulse up to 99 seconds (Fig. 10).

In contrast to the CW laser, the pulsed laser releases laser energy in short interrupted time intervals. The CW laser can deliver its wavelength in a "gated" manner of short bursts at relatively high powers (Fig. 11).

A second type of pulsed laser is the Q-switched laser which allows for maximum build-up of energy in the resonator and releases the energy on a shutter system within the head. The results are very short pulses of megawatt power. Besides having a thermal effect,

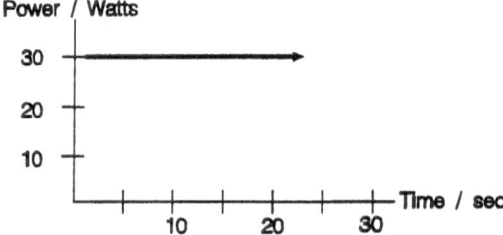

Fig. 10. Continous wave

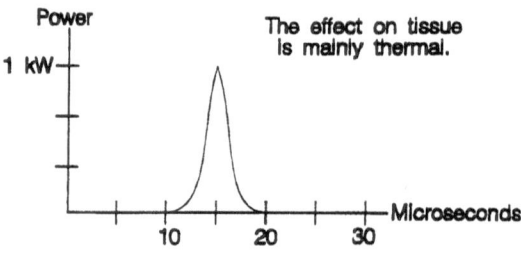

Fig. 11. Pulsed laser, short burst for less than a millisecond

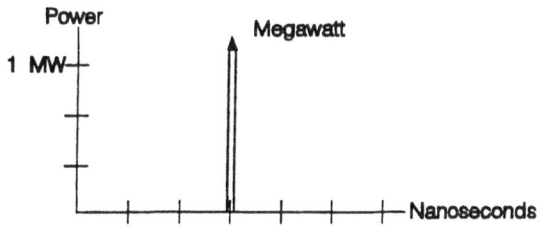

Fig. 12. Q-switched laser

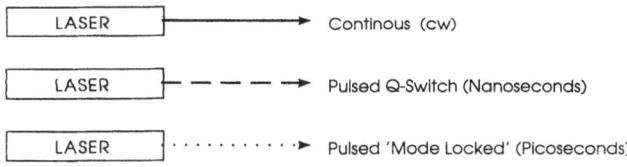

Fig. 13

the Q-switched beam has an impact of "shock wave" or acoustic effect on tissue (Fig. 12).

Two additional lasers in the category of "pulsed" are:

– "Normal mode pulse" (typically in the Nanosecond domain)

and

– "Mode-locked pulse" (typically in the Picosecond domain),

the energy output of pulsed lasers is expressed in millijoules.

In the ophthalmic laser system the Nd:YAG utilizes the "Q-switched" and "mode-locked" pulse.

In general: Variable pulse lengths are available (Fig. 13).

Opthalmic lasers work by giving high energy powers of short pulse durations to provide a mechanicalor shock impact to rupture membranes. The primary mechanism of damage occurs as the laser energy has an optical breakdown at or very near the focal point. Shock waves emanate from the optical breakdown and cause mechanical rupture of tissue in the area. The power density must be sufficiently high to cause this thermo-acoustic transient pressure wave.

Further application for the pulsed NdYAG laser lies in the destruction of stones in the gallbladder and ureter.

2. Laser-Tissue Interactions

The radiant energy of the laser beam can be transformed into heat energy that produces medical and surgical effects in tissue. A variety of factors combine to determine the nature and extent of this thermal effect: the power density or energy density, the spot size, and the nature of the light interaction with tissue.

The total power output of a laser is measured in watts. Of far greater importance to the physician, however, is the power density, measured in watts per square centimeter (watts/cm^2). Power density is directly correlated with the thermal effect induced in the target tissue, and determines the type of medically useful work, such as coagulation, vaporization, or cutting, that the laser will perform.

A related quantity, energy density, measured in power density times time, of joules per square centimeter, indicates the total amount of energy put into a given tissue area during the course of laser irradiation.

A second important concept is spot size, a measure of the surface area upon which laser light is concentrated. The spot size partially determines the extent to which thermal effects can be localized to perform precise medical procedures without affecting adjacent tissue. Spot size is also an important determining factor of power density. High power densities can be achieved by turning up the output power, but since density varies as the inverse of the area, one increases power density more dramatically by reducing the spot size. At a constant power, halving the spot size will quadruple the power density. One's ability to regulate the laser spot size depends upon the wavelength of the light and the means of delivery and focusing. A further concept governing the medical utility of lasers is the nature of light interaction with tissue. Light can be reflected from, transmitted through, scattered within, or absorbed by tissue. Only if the light energy is absorbed by tissue will it be transformed into effective thermal energy. If it is scattered, it will ultimately be absorbed over a larger tissue volume and produce a less intense, less accurately defined thermal effect. If reflected from or transmitted through tissue, it has no effect at all. The way in which light interacts with a substance largely depends upon its wavelength. Just as we commonly think of sunlight as being absorbed by black objects, passing through glass, and reflecting from white surfaces, monochromatic laser light selectively interacts with molecules of different "colours" (Fig. 14).

As the laser energy is absorbed by the tissue several surgical effects take place. At 60° C protein denaturation in the cell takes place and coagulation of blood vessels. At temperatures near 100° C the intracellular

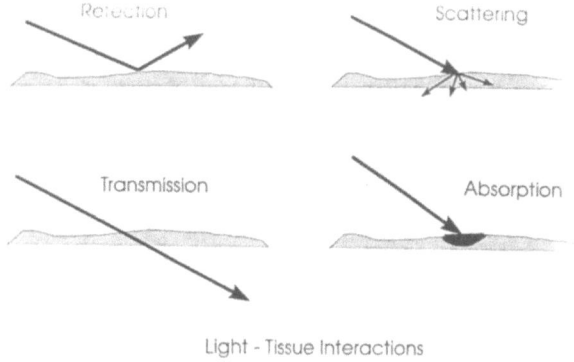

Light - Tissue Interactions

Fig. 14

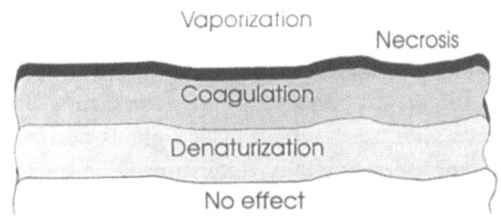

Fig. 15. Non-contact ND:YAG energy can cause 3–5 mm of thermal damage. Nd:YAG energy delivered by the contact sapphire will be limited to less than one millimeter of damage

Table 1. *Specifications of Surgical Lasers*

	CO_2	Argon	Nd:YAG
Type	gas	gas	chrystal
Wavelength	10.6 μm	488–514 nm	1,06 μm
Colour	far infrared	blue-green	near infrared
Power	0–100 W	0–20 W	0–150 W
Penetration	0,2 mm	1 mm	5 mm
Absorbed by	water	hemoglobin and melanin	tissue protein
Delivery	articulated arm	optical fiber	optical fiber
Aiming beam	He-Ne	argon	He-Ne
Energy efficiency	20 %	0,1 %	2 %
Maintenance	moderate	high	low
Coagulation	poor	fair	good
Cutting	good	poor	fair

The NdYAG wavelength is absorbed by the protein content of tissue and by dark pigmented tissue such as blood.

water begins to evaporate causing shrinkage and tissue loss. As temperatures increase beyond this point vapourization will occur. The cell content is turned to vapor. Below the vaporized tissue will be a shallow zone of necrosis (Fig. 15).

In summary, the combination of the laser's output power and the beam's spot size on the tissue determines the power density and distribution when the light reaches the target site. The combination of laser wavelength and the absorption characteristics of different tissues determines the three dimensional thermal response that will be induced. By manipulating these factors one can obtain the desired medical and surgical effects of cutting, coagulation, vaporization, or the delivery of homogeneous low levels of radiation over a broad area (interstitial irradiation).

3. Surgical Lasers

There are many advantages in the use of lasers in surgery. Lasers produce a sterile incision, and reduce both bleeding and operating time, especially when cutting highly vascular tissue. The ability to pass laser beams through rigid endoscopes or small diameter flexible optical fibers makes them well suited to endo-

scopic procedures, and broadens the range of treatment that can be performed on a minimally invasive, same-day basis. Following laser surgery healing is rapid, postoperative pain is reduced, and swelling and scarring are less than with other techniques. Under appropriate conditions lasers cannot only cut, but will coagulate or vaporize. Certain wavelengths of laser light are absorbed only by particular tissue components, allowing the selective destruction of targeted areas even when they are deeply embedded in healthy tissue (Table 1).

a. Carbon Dioxide Laser

The CO_2-laser emits light at 10.6 μ, in the far infrared region of the spectrum. Light of this wavelength is strongly absorbed by water with minimal scattering. Since most tissue is more than 80 % water, virtually all CO_2-laser energy is absorbed within 200 μ of the tissue surface. This creates a very localized and intense thermal effect, making the CO_2-laser a good cutting instrument that produces a sharp incision with minimal lateral necrosis. The CO_2-laser converts electrical energy to laser beam energy with relative efficiency, and comparatively small units can be used for many applications.

CO_2 energy exits the resonator in a collimated beam (wavelengths equal distance apart) yet some divergence takes place. The wavelength is passed through a lens either in the focusing hand-held device or microscope which converges the beam to the optimum spot size (Fig. 16).

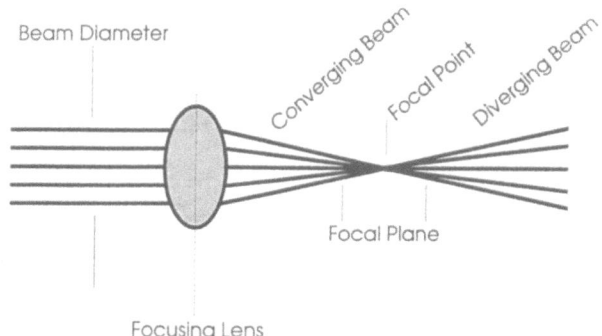

Fig. 16

Optimum spot size control offers greater control of power density. The CO_2 laser is able to focus a spot size as small as 30 microns in diameter. CO_2-lasers may be in the CW or "pulsed" modes. The beam is always delivered off the tissue surface, never in contact.

CO_2 energy is delivered to tissue by a series of perfectly aligned mirrors located in the "knuckles" of the articulating arm.

The articulating arm may terminate in a focusing handpiece or through an operating microscope. Because of the very long wavelength (10.600 nm), CO_2-energy cannot transmit through a flexible fiber. The wavelength could become distorted. Using the CO_2-beam in dyscoscopy and fluid hysteroscopy is impossible because of the moderate absorption by water.

Since infrared light is invisible, the CO_2-laser contains a secondary low power helium-neon laser to provide a visible aiming beam. The He Ne beam is aligned perfectly with the CO_2 beam allowing the surgeon to direct the energy accurately.

b. Argon Laser

Medical argon lasers emit a beam of light between 488–514 nm, in the blue-green range of the visible spectrum. Light of this wavelength passes through water and clear media with very little effect.

The argon beam scatters more than the CO_2-beam upon impact, and is primarily absorbed by tissue pigments such as haemoglobin and melanin, with the result that it more diffusely penetrates up to a millimeter of tissue and produces a less intense thermal effect.

For these reasons argon lasers are widely used in ophthalmological and dermatological applications, where their energy can be addressed through fluids or surface tissue to underlying pigmented target structures. The argon laser can only coagulate small vessels; penetration is too shallow for effective haemostasis of larger vessels, and in areas of active bleeding it creates a coagulated cap on the surface that prevents the beam from reaching the bleeding site beneath.

The argon beam passes through optical fibers, and hence can be used through flexible endoscopes. Moreover, the fact that the light falls within the visible spectrum obviates the need for a secondary aiming beam. In open surgical applications, the beam can be focused to small spot sizes with a handpiece, though the practical advantages of this are limited by the beam scattering following impact. The major drawback of the argon laser is its extreme inefficiency and consequent large size. Whereas a CO_2-laser can convert 100 watts of electrical power into 20 watts of beam output, an argon laser uses 20.000 watts of electricity to achieve the same result, which requires expensive, high current, three-phase wiring. Moreover, the excess energy creates tremendous heat that must be removed by a special pressured water cooling system. Even 20 watt argon lasers are massive machines that need to be permanently installed in the hospital. In addition, all argon lasers employ closed plasma tubes that must periodically be replaced at considerable expense.

Because of its limited power, beam scattering, and selective absorption by tissue pigments, the argon laser cannot deliver the power densities needed for cutting and vaporization. These factors, as well as its limited coagulating properties, effectively restrict its use to the types of specialty applications mentioned above, and make the argon laser a poor choice as a general surgical instrument.

c. Nd: YAG Laser

The neodymium: YAG is a solid state laser that uses a crystal of yttrium aluminium garnet (YAG)—a material widely used to make artificial diamonds—doped with a trace quantity of the rare earth neodymium, which serves as the actual lasing medium. The YAG wavelength most commonly used is at 1064 nm, in the near infrared region of the spectrum. As with the CO_2-laser, YAG light is invisible and a helium-neon aiming beam is required. Unlike the CO_2-laser, it is delivered through all flexible fiberoptic endoscopes. It will also pass through water with little effect, and can therefore be used in the bladder and other fluid filled cavities.

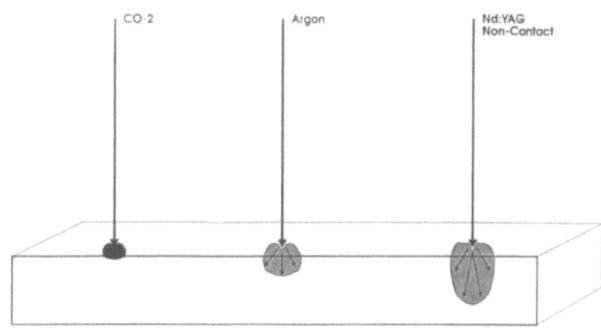

Fig. 17. Beam penetration and scattering in tissue

Table 2. *Comparison of Surgical Laser Features*

	CO_2	Argon/KTP	Nd:YAG
Beam easily deliverable to any part of the body	no	yes	yes
Beam transmissible through fluids	no	yes	yes
Convenience in providing any required power density	moderate	low	moderate
Precisely controllable focal point	no	no	no
Degree of damage to healthy tissue	low	moderate	high
Smoke generation	high	moderate	high
Tactile feedback	no	no	no
Laser power requirements	high	high	high
Laser maintenance requirements	moderate	high	low
Delivery system maintenance requirements	moderate	high	high

YAG light is diffusely absorbed by all protein molecules, and can pass through and affect tissue to a depth of three to five millimeters. This means that the Nd:YAG delivers a broad, deep, non-specific thermal effect that makes it an exceptional coagulator, even with large diameter vessels or at sites of active bleeding (Fig. 17).

The chief drawback of the YAG laser is related to light scattering. Whereas the CO_2-beam is absorbed almost immediately after impact, and the argon beam scatters somewhat, the YAG beam is highly scattered both within the tissue and away from the treatment site. This phenomenon contributes to the YAG beam's coagulative properties, but it also means that once the beam strikes tissue, 30–40 % of its energy is lost through backscatter. The remaining forward scattering laser light is spread through a considerable tissue volume. The YAG has sufficient power to cut and vaporize, but in doing so it causes more damage to adjacent healthy tissue than a CO_2-laser used in the same situation. In addition, the amount of backscatter depends to a large degree upon the angle of beam incidence, so that the power density delivered by the delivery system is held.

YAG lasers are extremely reliable. The Nd:YAG crystal seldom needs replacement, and is inexpensive compared to a plasma tube. The Nd:YAG lasers used in conventional, noncontact procedures need to deliver output powers of 100 watts or more, and these high-power units do require plumbing hook-ups and hard wiring. For Contact Laser Surgery, however, much lower power levels are used, and portable Nd:YAG lasers with completely self-contained cooling are sufficient for all applications. The advantages and disadvantages of the different surgical laser are listed in Table 2.

4. Contact Laser Probes

Contact Laser Probes offer a completely new method of delivering Nd:YAG energy to tissue, and overcome many of the limitations encountered with the three conventional noncontact medical laser systems. Using less that 25 watts of power Contact Laser Probes will cut, coagulate, vaporize, or administer low levels of interstitial irradiation. They can be attached to a variety of handles for use in open surgical procedures, or can be affixed to a standard optical fiber and passed through any rigid or flexible endoscope for use in an ever-increasing range of endoscopic applications.

Contact Laser Probes are made of a specially selected, physiologically neutral synthetic sapphire crystal with great mechanical strength, low thermal conductivity, and a high melting temperature (2030–2050° C).

They are used in direct contact with tissue, which allows precisely controlled manipulations and restores the tactile feedback that was lost in conventional laser techniques.

Due to the optical properties and geometrical design of each probe, Contact Laser Probes shape the power density to deliver the optimal laser energy intensity and distribution for each type of procedure. By selecting the appropriate probe and laser power one can not only determine the precise spot size and power density, but can control the shape and volume of thermal effect. This is not possible with any noncontact laser or with other thermal techniques (Table 3).

Table 3. *Comparison of Power Densities Contact vs. Non-Contact Nd:YAG.* Laser power (watts) required to achieve given power density at the tissue surface. Assumes typical non-contact Nd:YAG spot size of 1.5 mm and 0.4 mm diameter contact laser Scalpel. Power indicated is that detected after exciting fiber optic

Contact Nd:YAG	Non-contact Nd:YAG	Power density (watts/cm^2)
1	12	700
5	62	3.500
10	124	7.000
15	185	10.500
20	247	14.000
25	309	17.500

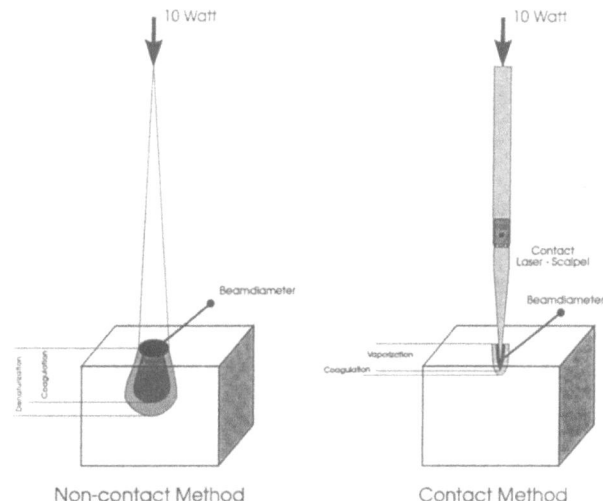

Fig. 18. Contact versus non contact

A conventional, non-contact YAG laser delivery system emits a diverging beam of gradually increasing size and diminishing power density. 30–40 % of the beam energy can be lost to backscatter, and a portion of the rest is expended on non-targeted healthy tissue due to inaccurate focusing and beam scattering. Contact Laser Probes create a well-defined localized region of high power density right at the tip of the probe, which is placed precisely against the target tissue. The problem of focusing is completely eliminated. Spot size and power density are under accurate and constant control. Energy loss to backscatter is cut to less than 5 %, and the overall laser output power needed to achieve a given therapeutic effect is 75–90 % lower than would be needed in a non-contact procedure. Since less total energy is delivered to the site, damage to healthy neighbouring tissue is greatly reduced. (Fig. 18). The Nd:YAG laser is ideally suited for use with Contact Laser Probes. The YAG is not limited by its absorption spectrum to any particular tissue type. It is the best available thermal coagulator, has good penetration, and its drawbacks

in the non-contact mode-poor cutting, excessive backscatter, inaccuracy, and excessive tissue damage—are precisely eliminated by Contact Laser Probes. Since 25 watts is the maximum power used with Contact Laser Probes in any therapeutic situation, a compact, low cost Nd:YAG unit will meet every procedural requirement. A single portable laser system can be used in inpatient or outpatient operating rooms, the endoscopy-suite, or the physician's office, providing an extremely versatile and cost-effective addition to any hospital or clinic (Table 4).

Both experimental and clinical studies with Contact Laser Probes have produced remarkable results. Resection of rat liver could be accomplished at 5 watts with a Contact Laser Scalpel, whereas non-contact YAG powers below 20 watts invariably resulted in fatal bleeding complications. The resection was performed much more quickly with the low power con-

Table 4. *Comparison of Surgical Laser System Features*

	CO$_2$	Argon/KTP	Non-contact Nd:YAG	Contact Nd:YAG
Beam easily deliverable to any part of the body	no	yes	yes	yes
Beam transmissible through fluids	no	yes	yes	yes
Convenience in providing any required power density	moderate	low	moderate	high
Precisely controllable focal point	no	no	no	yes
Degree of damage to healthy tissue	low	moderate	high	low
Smoke generation	high	moderate	high	low
Tactile feedback	no	no	no	yes
Laser power requirements	high	high	high	low
Laser maintenance requirements	moderate	high	low	low
Delivery system maintenance requirements	moderate	high	high	low

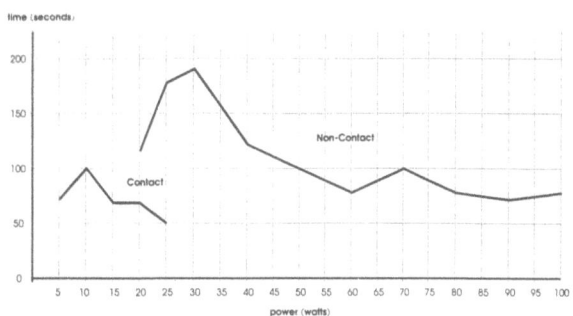

Diagram 1. Time required for rat liver resection as function of laser power

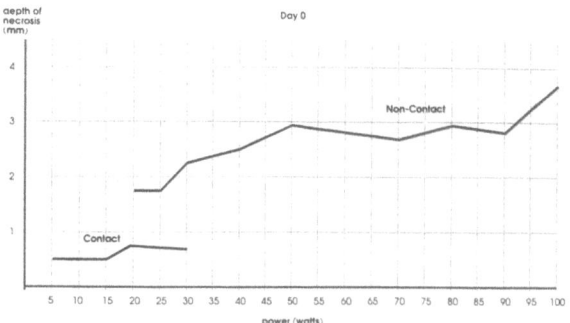

Diagram 3. Depth of tissue necrosis as function of laser power

tact probe than was possible with a non-contact YAG beam (Diagram 1).

Blood loss was significantly less in the contact surgery group because of the Laser Scalpel's unique ability to control the balance between cutting and coagulation. The non-contact YAG beam produced excessive smoke at high powers that required evacuation, whereas smoke was not a problem at any of the powers needed in the contact situation.

Most significant, however, is the difference in tissue necrosis with the two techniques. The non-contact resection caused up to 3 mm of lateral necrosis, depending upon the laser power. The contact procedure, on the other hand, did not cause liver necrosis of more than 0,5 mm under any circumstances. This was true both immediately post-operative and at 15 days after surgery, implying more rapid healing following contact laser surgery (Diagram 2).

In clinical application Contact Laser Probes demonstrate parallel advantages: contact procedures are better controlled and less traumatic than conventional non-contact laser techniques. In endoscopic procedures in particular, Contact Laser Probes have proved to be dramatically effective, and so versatile

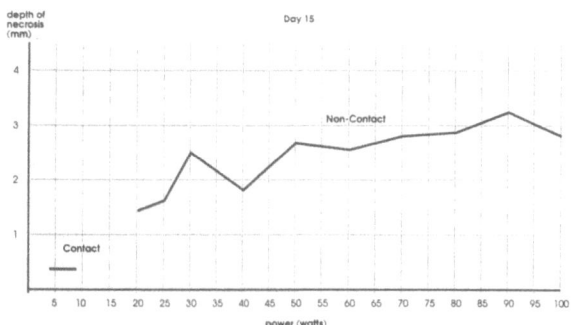

Diagram 4. Depth of tissue necrosis as function of laser power

that they lend themselves to treatments never possible with non-contact lasers or with any other techniques. Damage to adjacent healthy tissue is reduced by up to 75 %, and there is consequently much less sloughing of necrotic tissue and a lesser incidence of subsequent infection. The depth of the thermal effect is carefully controlled, which lowers the risk of perforation when working in hollow viscera, and reduces pain since the thermal effects do not penetrate through to the serosal surface (Diagrams 3 + 4).

5. Principles of Contact Laser Surgery in Endoscopy

SLT Contact Laser Probes are designed to serve four general medical functions: cutting, vaporization, coagulation, and delivery of interstitial irradiation. Each of these functions results from the induction of a thermal effect of given intensity in a certain tissue volume. Cutting requires intense, highly localized heat to vaporize small tissue volumes rapidly, creating a controlled incision with little damage to adjacent areas. For vaporization of fairly large tissue volumes, an intense but broader thermal effect is needed. Coag-

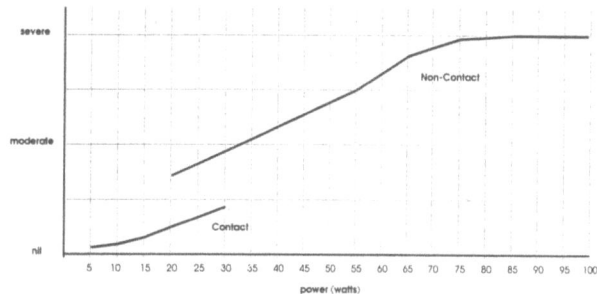

Diagram 2. Amount of smoke as function of laser power

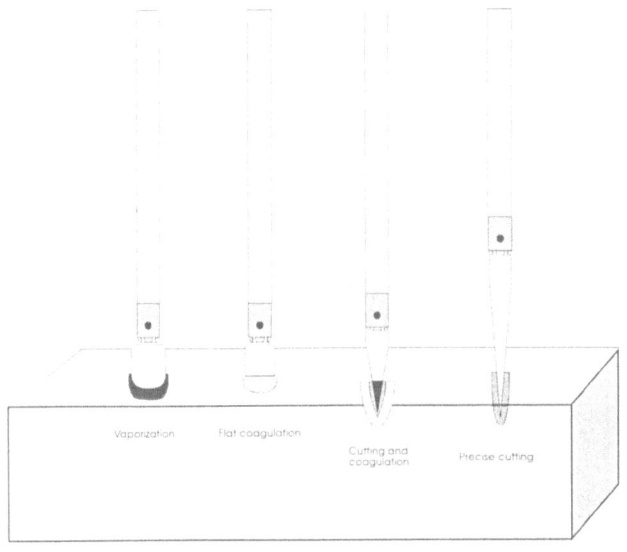

Contact laser probes

Fig. 19. Contact laser probes

ulation requires milder temperatures in yet larger volumes. For interstitial irradiation, where a probe is inserted into the tissue, one needs a fairly mild effect that is homogeneously distributed laterally as well as along the distal axis.

Contact Laser Probes are designed in a variety of configurations that produce power density distributions well suited to these specific therapeutic tasks (Fig. 19). The synthetic sapphire probes are available in two formats: there is a set of large incising probes, called Laser Scalpels, used exclusively as handheld instruments in open surgery; and there are various small Contact Laser Probes that are predominantly used in endoscopic applications but which may, if desired, also be used in open procedures. All Contact Laser Probes can be used with any Nd:YAG laser that is stable at low output powers.

The selection of interchangeable Laser Scalpels have distal tip diameters ranging from 0.2 to 1.2 mm, and screw into a variety of sterile disposable handles. Laser Scalpels are used with the Nd:YAG laser at powers below 25 watts, and in some applications at powers as low as 2–3 watts. It transforms the Nd:YAG into a multi-disciplinary, multi-purpose surgical laser that cuts as cleanly and precisely as the CO_2-laser yet retains the coagulative properties of the non-contact YAG.

Contact Laser Probes are designed primarily for use in endoscopic procedures. They screw onto a met-

al Universal Connector that fastens to the SFE 1.8 or 2.2 mm OD SLT disposable quartz fibers, and will pass through rigid endoscopes or the biopsy channel of any standard flexible fiberoptic endoscope. Different probe geometries shape the laser energy distribution into appropriate configurations for vaporization, coagulation, cutting, or interstitial irradiation. The user can, if desired, change probes during a procedure simply by withdrawing the fiber from the endoscope, unscrewing one probe, and screwing on another.

SLT factory mounts each universal connector on the disposable fiber for optimal transfer of energy. Unfortunately other laser companies are instructing hospital personnel to mount the connectors on already existing fibers. The room for error is great when not applied in a factory setting, leading to potential problems in energy delivery.

Contact Laser Probes are used with the Nd:YAG beam is effective in water or at sites of active bleeding, and provides exceptional coagulative abilities. Contact Laser Probes give complete control of energy penetration and dispersion, minimizing unwanted damage to healthy tissue. Since very low laser powers are used, there is minimal backscatter and virtually no smoke. Significantly less heat is generated around the target tissue. Contact Laser Probes are designed to be water cooling, eliminating the discomfort and dangers of gaseous distension of the patient.

Due to the physical characteristics of the synthetic sapphire crystal, Contact Laser Probes can be used in direct contact with tissue or blood without danger of melting. Once the Universal Connector has been affixed to the fiber, there is no need for subsequent recleaning or polishing. The direct contact technique makes possible more accurate targeting of the laser effect, even on moving tissues, and permits coaptation of bleeding vessels for simpler and more rapid haemostasis. The Laser Scalpel and Contact Laser Probes are safe and reliable, but a few important points should be kept in mind:

Since very low laser output powers are transformed into high power densities at the tissue, any Nd:YAG laser used with them must provide a stable power delivery over the entire range.

Heating is much less of a problem than with non-contact laser irradiation, but coaxial gas or water cooling is still necessary in most situations to cool the probe and clear the surgical site.

The Laser Scalpel and Contact Laser Probes should

Fig. 20. SLT 60W laser system

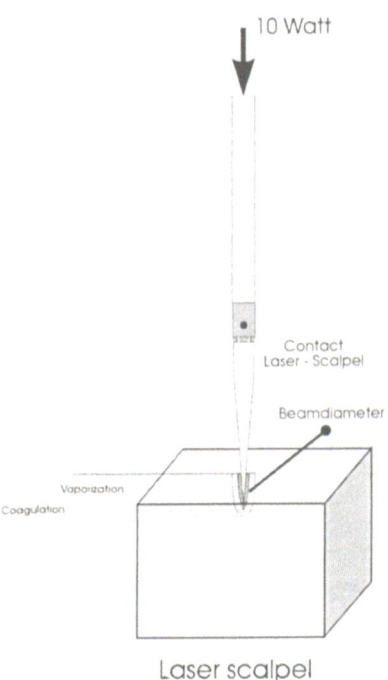

Fig. 21. Contact laser scalpel

be placed in contact with tissue before power is turned on. Power should be kept on momentarily while disengaging from tissue, as this eliminates the chance of tissue sticking.

If power is kept on for more than about five seconds while the Scalpel or Probes are not in contact with tissue, the sapphire tip may turn white and change shape. Deformation of the probes alters the pattern of power density distribution, and melted probes must be replaced.

In many ways Contact Laser Probes provide a safer means of laser energy delivery than conventional non-contact systems, but users should nonetheless follow all precautions recommended for the particular laser being used (see Fig. 20).

6. Applications in Endoscopic Neurosurgery

1. Cutting

SLT incision probes provide simultaneous cutting and coagulation, while causing minimal damage to adjacent healthy tissue. Cutting can be accomplished with any of the Laser Scalpels in open surgery or with the Chisel and Conical Probes in endoscopic procedures (Fig. 19). The geometric and optical properties of the synthetic sapphire probes are such that the YAG beam is brought to a tight focus and very high

power density exactly at the tip of the probe. Power density then drops off rapidly at short distances from the tip, giving an extremely localized and accurately controlled thermal effect. In other words, "What you see is what you get." Cutting takes place at the tip, not in adjacent tissue. When the probe is removed from the tissue, cutting ceases. This is not the case with the non-contact YAG beam. Moreover, since the thermal effects in contact cutting are so carefully circumscribed, the invisible subsurface tissue damage that can occur in non-contact YAG procedures is reduced by over 80 % (Fig. 21).

Laser Scalpel

The diameter of the distal tip of the Laser Scalpel defines the spot size of the laser beam when the Scalpel is in contact with tissue. Different Laser Scalpel tip diameters give the surgeon the ability to select either greater cutting or greater coagulation. The choice of scalpels depends upon the tissue texture and vascularity. Smaller tip diameters, for example, can be used for an initial clean incision, whereas when working with soft, highly vascular tissue such as the liver or spleen, larger tip diameters provide better haemostasis. The maximum power required is 25 watts, though precise cutting can be achieved at substantially lower power levels. In general, the smaller the tip diameter, the lower the power level needed to produce a given tissue effect. The Laser Scalpels screw easily onto the various handles, and the surgeon can quickly change tips as circumstances suggest.

While holding the tissue taut, the Laser Scalpel should be drawn lightly across the surface of tissue, not used mechanically to separate or tear. It is the laser energy rather than physical pressure of the probe that is creating the incision. Incision depth is determined by a combination of tip diameter, laser power, and the speed of movement across the tissue.

The Laser Scalpel should be held in a comfortable and natural position, much as one would hold a pencil, though at a slightly steeper angle to the tissue surface. Since the Laser Scalpel crystal channels beam energy directly along its longitudinal axis, with virtually no lateral irradiation, the cutting effect is somewhat greater as the orientation approaches perpendicular.

At the initial incision, the laser Scalpel may appear to cut more slowly than a steel scalpel or the CO_2-beam. Over the course of an entire procedure, however, the Laser Scalpel is considerably quicker since it efficiently cauterizes as it cuts. When working with highly vascular tissue, there is no need to stop to ligate or clip most bleeding vessels. Vessels as large as 2 mm in diameter can be cut and sealed simply by moving the Laser Scalpel slowly across them. If necessary, one can go back and coagulate bleeding spots without changing instruments by dabbing lightly at the area with the scalpel. For larger vessels, light dabbing or stroking with the Laser Scalpel along each side of the vessel will create oedema, after which the vessel can be cut.

The Laser Scalpel uses power levels 75–90 % lower than those needed by non-contact lasers to accomplish a given surgical effect. There is minimal backscatter, a constant and well-defined spot size, and less energy is put into the tissue. The result is far less damage to adjacent healthy tissue, as seen from histological studies of incision sites. Operating time and bloodloss are greatly reduced, resulting in an overall decrease in post-operative complications.

The different Contact Laser Probes offer a choice of endoscopic cutting styles. Powers below 15 watts are sufficient for most therapeutic situations, and some operations can be performed at much lower powers. Users should realize, however, that recommended power settings are of less importance than the actual observed tissue effects, and changes in the tissue texture and colour are the only reliable indications of the laser effect.

The Conical Probe provides a power density distribution pattern very similar to that of the Laser Scalpel and is used in much the same way to create fine incisions. Since the tip of the Conical Probe should ideally be moved laterally across the tissue surface rather than pressed against it, the use of a rigid endoscope offers certain advantages, since once the probe has been oriented against the tissue, the entire delivery system can be gently shifted to guide the incision.

The Chisel Probe will cut directly through tissue, though its primary use, as its shape and name suggest, is to shave off thin layers of tissue. The Chisel Probe vaporizes and coagulates as it cuts, providing a clean and bloodless site. Used with a rigid or flexible fiberoptic endoscope, this probe can be pressed to the tissue surface and pushed lightly across it. It is ideal for recanalization of a totally or partially obstructing carcinoma of the oesophagus, bronchus, or colon. It removes tissue more rapidly than vaporization with the Rounded Probe, and is effective for quick preliminary work when there is no danger of perforation. To achieve optimal cutting and greatest depth control, the Chisel Probe should be held with on the flat nearly parallel to the tissue.

2. Coagulation

Hemostasis is accomplished by heating tissue sufficiently to produce oedema and protein coagulation. For this purpose one needs a power density distribution that creates a milder, broader, and deeper thermal effect than is required for cutting or vaporization. The Nd:YAG laser is generally the instrument of choice for coagulation because its beam penetrates fluid and tissue to a depth sufficient to seal, rather than merely cap, bleeding vessels.

The non-contact YAG requires 60–100 watts of power for coagulation, however, which causes a certain degree of uncontrolled tissue damage and occasionally precipitates severe bleeding as a result. Contact Laser Probes, on the other hand, safely and effectively seal vessels up to 3 mm in diameter using powers below 10 watts. All Contact Laser Probes provide a certain measure of coagulative ability, but the physical shape and power density-distributions of the Flat and Frosted Probes are best suited to this purpose.

The Flat Probe is the instrument most often used for general coagulation. At power settings between 8 and 10 watts it provides much more rapid haemostasis than other laser or thermal methods with minimal necrosis in adjacent tissue. The technique of providing haemostasis with the Flat Probes is similar to that used with a non-contact laser. Rosettes are formed around the periphery of the bleeding vessel to initiate oedema. After the bleeding visibly decreases, one can use the probe mechanically to coapt the vessel and seal it with short pulses of energy.

By setting the laser for 2–3 second intervals, setting the probe down, pressing the footpedal, lifting off the probe and releasing the footpedal, one establishes a simple repetitive pattern for rosette formation that eliminates the risks of tissue sticking and of melting the probe by keeping power on while disengaged from tissue. Alternatively one can set the laser in the continuous mode, hold the footpedal down, and walk the probe around the bleeding site, touching the tissue for half second intervals in each spot. As in all applications, the durations and power settings will vary slightly according to the procedure and tissue blanching that occurs when coagulation takes place.

The Frosted Probe, which was designed primarily for use in interstitial irradiation, can also be used for deep coagulation. By pressing this probe into mucosal tissue up to its flange, and using powers below 10 watts with short pulse durations, it provides effective coagulation of vessels several millimeters below the tissue surface. Because the power is low and the probe is embedded in the tissue, neither air nor water cooling is needed with this technique.

3. Vaporization

Vaporization requires higher power density and temperature than coagulation, and a broader energy delivery than is needed for cutting.

The Rounded Probe allows rapid removal of thin layers of tissue for controlled debulking of large tissue volumes. The Chisel Probe, as described in the cutting section, can also be used for a cruder and quicker combination of shaving and vaporization in areas where there is little risk of perforation.

When the Nd:YAG laser is used in the non-contact mode, vaporization is generally performed at 70–100 watts and produces a mixed effect of coagulation and vaporization with significant bleeding and subsurface tissue damage. In endoscopic applications,

where the precise distance of the fiber from the target tissue and the angle of beam incidence are difficult to control, the power density and thermal effect at and below the tissue surface vary greatly and unpredictably. The Rounded Probe, on the other hand, delivers approximately the same power density as an 80 watt non-contact YAG beam using only 15 watts of power, and produces true vaporization with minimal subsurface effects yet with sufficient coagulation to limit bleeding. Vaporization takes place only when the probe is in direct contact with the tissue.

There are many advantages to the true vaporization delivered by the Rounded Probe. For endoscopic oesophageal cancer removal, for example, the conventional non-contact technique involves coagulating to some depth below the surface, causing necrosis, waiting 48–72 hours for the blanched layer to slough off, and then repeating the process as necessary until the tumour has been removed. Until sloughing has taken place it is not clear how deep tissue damage extends, and one must therefore proceed cautiously, layer by layer, over the course of three to eight treatment sessions. The results still may not be altogether satisfacory, since the procedure is limited by the danger of thermal or necrotic perforation.

The Round Probe, on the other hand, actually removes layers of cells, and can vaporize tissue to an arbitrary depth in the course of one or two sessions. The probe is simply dabbed at or stroked across the target site, and vaporization is indicated by a slight darkening of the tissue. Because the power density is localized at the rounded surface of the probe, there is less danger of immediate thermal perforation or of later complications due to necrosis of underlying tissue, and one can confidently work closer to the mucosal wall.

Water delivery is sufficient to cool the probe during endoscopic vaporization, which eliminates the discomfort and dangers of patient distension, and further reduces the small quantities of smoke generated by a debulking procedure. Since there is less uncontrolled heating of deep tissue layers, contact vaporization is also less painful to the patient.

4. Interstitial Irradiation

The primary present function of the Frosted Probe is to provide deep coagulation, but its chief future applications will be in the administration of local hyperthermia and photodynamic therapy.

The Frosted Probe is pushed into the tissue up to its flange and delivers YAG power density and thermal energy in a hemispherical volume of 2–3 centimeters radius. With powers of 5–7 watts the Frosted Probe coagulates, but at powers in the 1–3 watt range it provides local hyperthermia. For the treatment of areas more than 2–3 cm in size, multiple probes can be implanted. This allows far more accurate control of treatment sites than current methods.

Correspondence: K.R. Goebel, Dipl. Ing., Surgical Laser Technologies, de la Fosse Weg 26, D-64289 Darmstadt, Federal Republic of Germany.

Acta Neurochir (1994) [Suppl] 61: 34–39

Magnetic Field Guided Endoscopic Dissection Through a Burr Hole May Avoid More Invasive Craniotomies
A Preliminary Report

K.H. Manwaring[1], M.L. Manwaring[2], and S.D. Moss[1]

[1] Phoenix Children's Hospital, Phoenix, Arizona and [2] Washington State University, Pullman, WA, U.S.A.

Summary

The neuroendoscope, coupled with radiofrequency or laser dissecting tools, can effectively resect obstructing membranes, biopsy and debulk tumor, and evacuate hematomas when the pathology is within the ventricular system. This less invasive approach through a burr hole usually avoids craniotomies. When the abnormal condition is within parenchyma or in the presence of opacifying bloody fluid, landmarks are not recognizable and the neurosurgeon quickly becomes disoriented. A more extensive craniotomy or a stereotaxic-guided procedure is then necessary.

We describe our preliminary experience with a geographic intracranial navigation system using realtime measurement of electromagnetic field strength in multiple planes to precisely indicate the position of the tip of the endoscope. A transmitting antenna is positioned beneath the patient's head. A 1.5 centimeter cubic antenna receiver is mounted upon a lenscope with instrument channel. The scope is guided into the surgical field after insertion through a burr hole. A square wave pulsed electromagnetic field measurement is made 140 times per second with correction for the earth's magnetic field once per second. Intracranial position data for the dissecting tip in regard to X, Y, Z, pitch, roll and yaw are output to a digitized computer map of the patient's MRI or CT scan. Also displayed on the computer screen is the video image from the endoscope. The neurosurgeon thus has simultaneous realtime geographic and near-field localization as he dissects. Electromagnetic field guided accuracy is within 2.0 mm inside the allowable 24 inch working sphere about the patient's head. Coupled with near-field video precision, accuracy is within 1 mm of recognizable dissection planes. Precise intracranial localization may make many pathologic conditions safely dissectable using the neuroendoscope, including parenchymal hematomas and neoplasms. Less invasiveness is expected to shorten hospitalization and patient morbidity.

Keywords: Endoscopic neurosurgery; magnetic field guidance; computer assisted surgery.

Introduction

Over the last decade, improved recognition, treatability and outcome in neurosurgical disease has evolved greatly due to enhanced imaging of brain anatomy and physiology. Computed tomography (CT), magnetic resonance imaging (MRI), and positron emission tomography (PET) have allowed, (with decreasing degree of geographic accuracy respectively), the improved localization of focal disease processes. Frame stereotaxy has facilitated importation of such accuracy into the surgical field with resulting decreased excisional and cranial exposure and more precise evacuation or excision of pathologic processes[1]. Reports of frameless stereotaxy are further promising due to the allowance during the surgical procedure of realtime guidance[6]. Such technology may decrease patient morbidity by showing the surgeon at craniotomy the location on image maps of his dissection, thus avoiding possible transgression of a vital structure or blood vessel.

During this same time, development and modification of existing lenscopes and fiberscopes for the intraventricular environment has resulted in effective techniques of fenestration, membranectomy, septostomy and biopsy[4]. We have developed and reported success with a saline radiofrequency dissector or "saline torch" which may vaporize tissue, allowing irrigation of the gases out through the endoscope burr hole[2,4,5]. However, principal limitations to more extensive dissection and tissue debulkment have been (1) bleeding and the adequate tools for achievement for hemostasis and, (2) maintenance of surgical orientation in the endoscopic near-field, particularly in the presence of opacifying bloody cerebrospinal fluid or tissue.

In an effort to extend present endoscopic capabilities of dissection, avoid excessive bleeding, and possibly avoid craniotomies in many neurosurgical pathologic processes, we developed hardware and software adaptations for a new technique of frameless stereo-

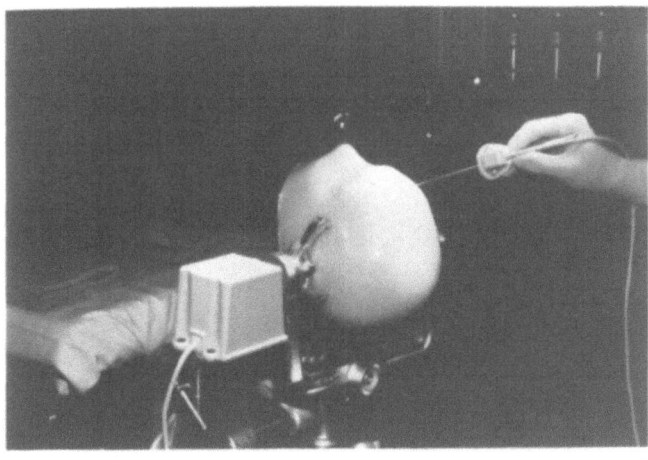 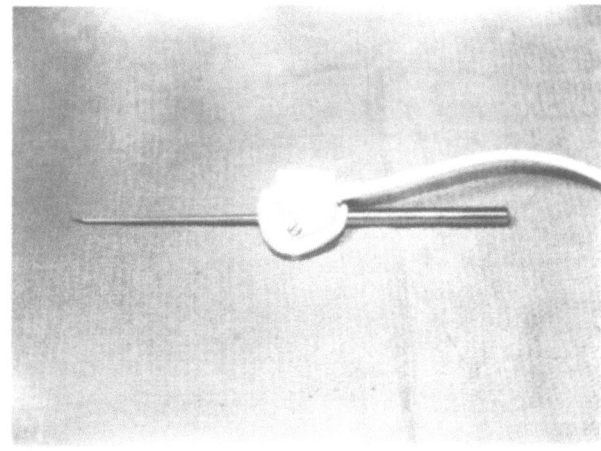

a   b

Fig. 1. (a) An antenna emitter is secured by a stainless steel post to the three point cranial fixation device thus maintaining a constant relationship between transmitter and surgical field. Alternatively, a horseshoe frame can be used, but careful attention must be given to assure that the head is not moved after translation of the magnetic field coordinate system into CT or MRI coordinates. Magnetic field receiver antenna may be attached sequentially to instruments in planning and performance of the surgery including Penfield No 4 dissector (b), marking pen and endoscope

taxy based on measurement of magnetic field strength. The receiver antenna is coupled with the lenscope or a guiding sheath for the fiberscope, providing the surgeon on the computer monitor simultaneous realtime video imaging from the endoscope camera and localization-orientation information upon the CT or MRI slice where he is operating. A similar magnetic field guidance system of frameless stereotaxy has been described by Kato[3].

Description of System

Our computer assisted endoscope guidance system consists of three components:

(1) magnetic field transmitter and receiver antennae with associated hardware,
(2) computer based frame grabber imaging boards and associated software and,
(3) endoscope coupling for the magnetic field receiving antenna*.

An electromagnetic field emitter antenna is secured by stainless steel mount with a three point cranial fixation device or horseshoe headrest (Fig. 1)**. The transmitter emits a continuously pulsed electromagnetic field in the allowable 0.6 meter (2 feet) operating sphere about the cranium. A receiver antenna is attached sequentially to instruments for incisional planning, burr hole preparation and endoscope guidance. Following calibration of the position of the tip of the marking pen or the endoscope, the position in milli-

meters in the transmitting antenna as well as orientation (pitch, roll, yaw) of the instrument are output through three dimensional digitalization hardware and software displayed on the screen. This data is displayed in the selected horizontal, sagittal, or coronal plane 140 times per second. The position of the tip of the instrument or endoscope viewing tip (X, Y, Z) with respect to the transmitter antenna $(O: X, Y, Z)$ is calculated using the following equation:

$$(X, Y, Z) = (a, b, c) + (\alpha, \beta, \gamma)$$

$$\begin{bmatrix} 1 & 0 & 0 \\ 0 & \cos R & \sin R \\ 0 & -\sin R & \cos R \end{bmatrix} \begin{bmatrix} \cos E & 0 & -\sin E \\ 0 & 1 & 0 \\ \sin E & 0 & \cos E \end{bmatrix} \begin{bmatrix} \cos A & \sin A & 0 \\ -\sin A & \cos A & 0 \\ 0 & 0 & 1 \end{bmatrix}$$

where a, b, and c are the coordinates of the magnetic field sensor determined by the 3-D digitizer (yaw, roll, and elevation, respectively); and α, β, and γ are the coordinates of the probe tip with respect to the magnetic field sensor reference frame $(P: U, V, W)$[3].

Translation of the magnetic field digitized position coordinate system to the imaging coordinates is performed as follows. Individual CT, MRI or PET images are imaged with a CCD video camera and stored to disk on the 8086 based personal computer through the frame grabber board. During radiographic imaging, a zero plane horizontal cut is determined by alignment of the slice angle to a band worn around the patient's head extending from the glabella and over the ears. The silicone tubing is barium impregnated and filled with oil allowing recognition on scout views of both CT and MRI. At induction of anesthesia, the receiving antenna coupled with the calibrated marking pen tip is used to trace the exposed cranium and scalp in the identical horizontal cut. The head is subsequently traced in a true sagittal midline from nose to external occipital protuberance (Fig. 2). The appropriate image slice for the horizontal cut is curve-fitted on the computer screen by eyeball alignment using a combination of zoom and rotation. The sagittal image is curve-fitted to the midline nose to inion tracing in similar fashion. Upon completion of this image translation, great care must be exercised not to allow the head to move in relationship to the transmitting antenna for the duration of the surgery.

Upon completion of head coordinate translation, the computer

* Prototype for intracranial navigation system, Codman & Shurtleff, Inc., Randolph, Massachusetts; NTSC plus video imaging boards, Redlake Corporation, Morgan Hill, California; endoscope mount by Southwest Orthotics, Inc., Phoenix, Arizona.

** Codman & Shurtleff, Inc., Randolph, Massachusetts.

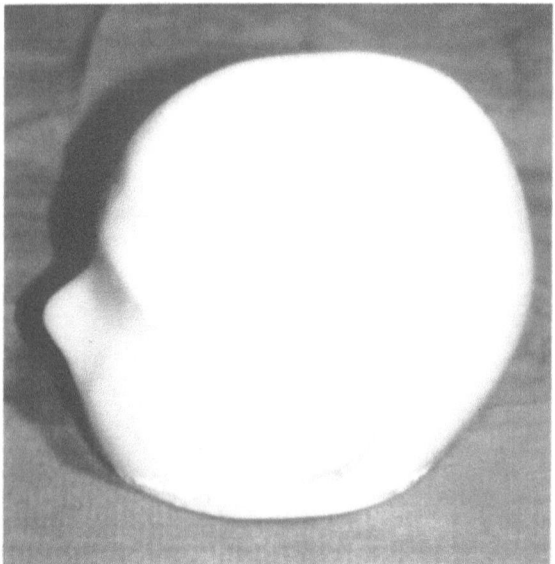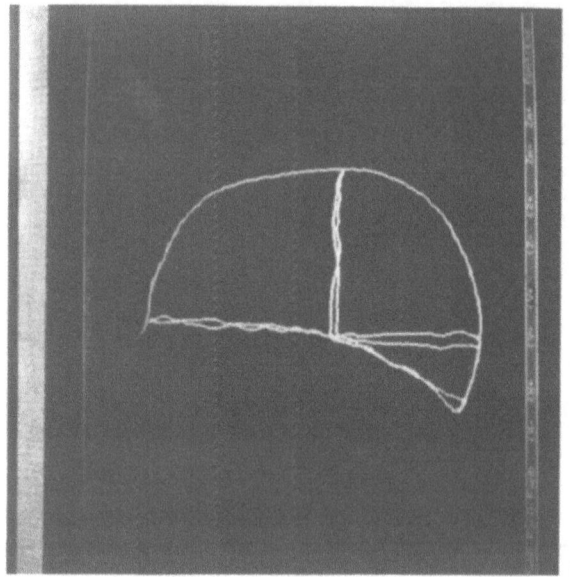

Fig. 2. (a) By tracing the head in the $Z = 0$ horizontal plane and curve fitting to the identical CT or MRI slice, the coordinate system of the magnetic field guidance system is translated into that of the radiographic image map. (b) The sagittal scout image is similarly aligned. If the transmitter-cranium relationship remains fixed, precise localization information on the surface and within the cranium will be displayed within the accuracy limits of the system. Here is seen the horizontal and sagittal scout CT cuts of a dummy head filled with plaster of paris and filler composite

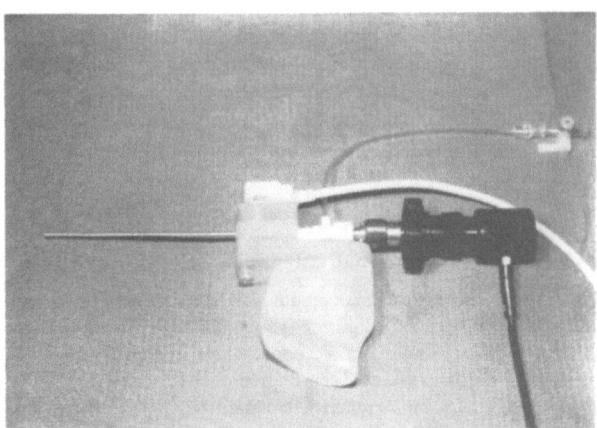

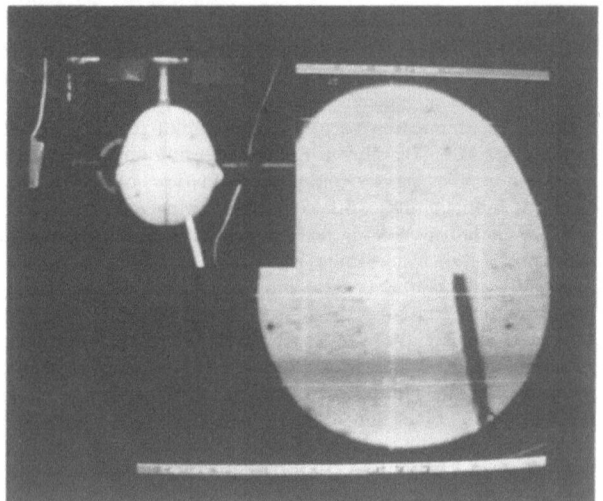

monitor now automatically brings to view the image slice in a selected horizontal, coronal or sagittal plane where the surgical instrument or endoscope touches the scalp, passes across the cranium, or penetrates the brain. After completion of selection of an incision using the attached marking pen, the receiver antenna is coupled to a zero degree lenscope with instrumentation channel***. A burr hole is created, the dura coagulated and incised and a peel-away sheath**** passed into the ventricular system, cyst, or parenchymal lesion. The coupled lenscope-receiving antenna is then passed into the peel-away sheath until its end is reached (Fig. 3). The endoscope realtime image is simultaneously displayed on the monitor screen along with the radiographic cut where the tip lies (Fig. 4).

*** Hopkins, Karl Storz, Pasadena, California.
**** Peel-away introducer, Codman & Shurtleff, Inc., Randolph, Massachusetts.

Fig. 3. Lenscope in its endoscope holder. The magnetic field receiving antenna is coupled to a lenscope holder. Light cord, video camera cord and its instruments connect behind the antenna mount
Fig. 4. The simultaneous display of the endoscopic image and the CT slice map of position within the ventricular system and orientation of the lenscope is presented to the surgeon. The monitor screen is positioned at the foot of the surgical bed

System Accuracy

Accuracy was tested in three modes: multiple millimeter rulers within the allowable .6 meter surgical sphere at a variety of orientations (Fig. 5); measurement of error between landmarks identifiable on CT of a dummy head (Fig. 6); and accuracy when coupled with the lenscope (Fig. 7).

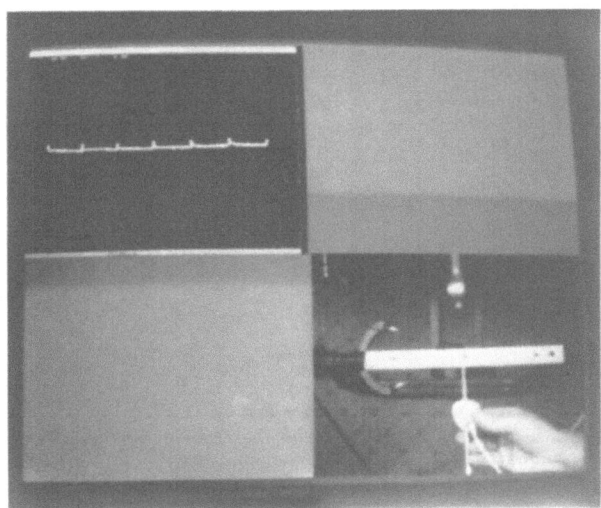

Fig. 5. Accuracy in 100 consecutive trials in measurement of 5 cm increment markers on a millimeter rule in space was consistently within 2 mm of the exact measurement. Here a computer screen tracing shows ticks at 0, 5, 10, 15, 20 cm and so-forth. In the straight axis, worse accuracy was .17 cm. This was irrespective of orientation

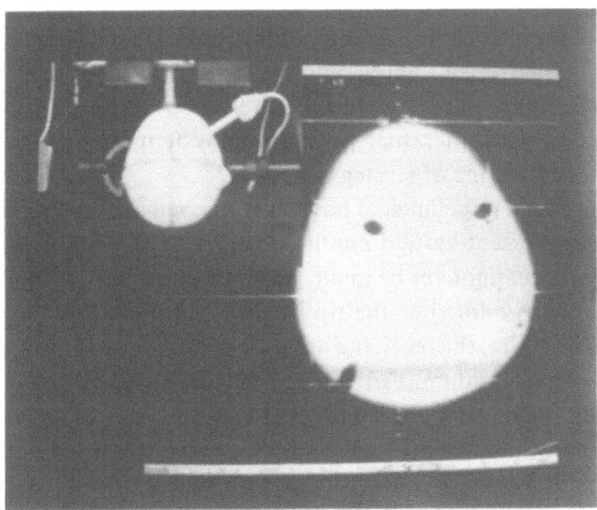

Fig. 6. A dummy head was prepared with multiple burr holes along typical trajectories for shunt tracks as well as with implantation in the surface of metallic spheres. The composite accuracy on 100 trials comparing CT scan slices and numeric output of the tip location of a Penfield No 4 dissector or marking pen was never worse than 5 mm (after curve fitting translation of the coordinate system)

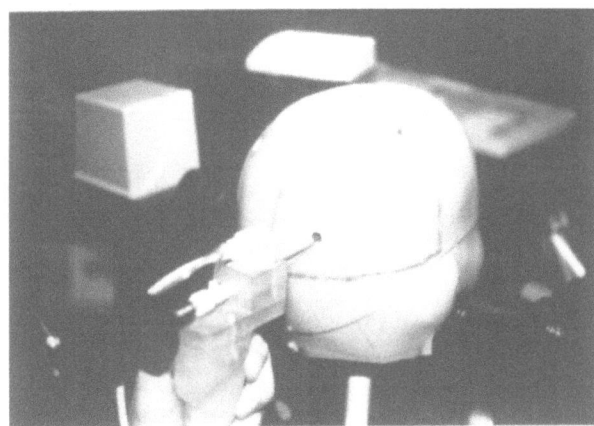

Fig. 7. With passage of the lenscope down burr hole trajectories, accuracy was always within 1 mm, i.e. worst error easily brought the target into the lenscope angle of view, allowing correction under direct endoscopic visualization

Clinical Trial

Our preliminary clinical trials to date have been limited to preoperative surgical planning and modeling for placement of the ventriculoperitoneal shunt and simple membranectomy with monopolar radiofrequency dissection. Figure 8 shows the simultaneous positioning appearance of the endoscope in the lateral ventricle and frontal horn appearance through the

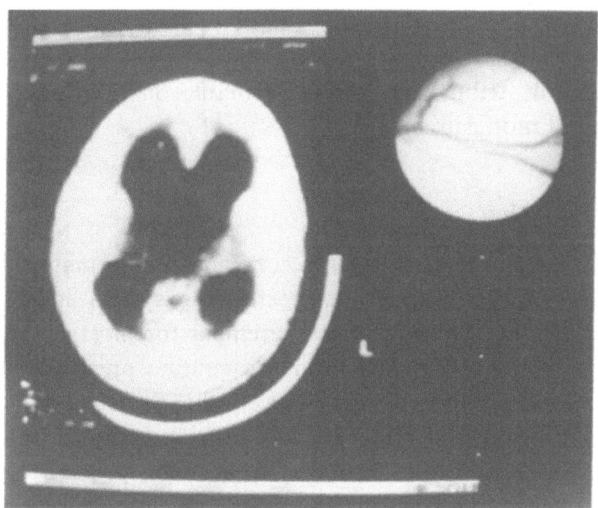

Fig. 8. From a typical occipital approach the tip of the endoscope can be passed toward the frontal horn for ideal positioning of a ventriculoperitoneal shunt. Alignment of the lenscope as passed from a posterior burr hole is demonstrated. A peel-away sheath is guided by the lenscope to a frontal horn position anterior to choroid plexus. A ventriculoperitoneal shunt may then be positioned in the exact selected location. The frontal horn typical appearance of ependymal veins is seen

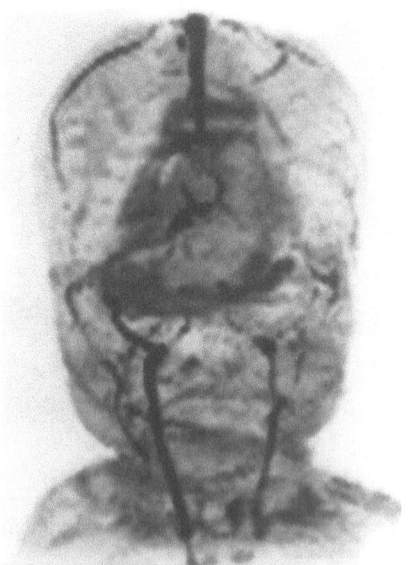
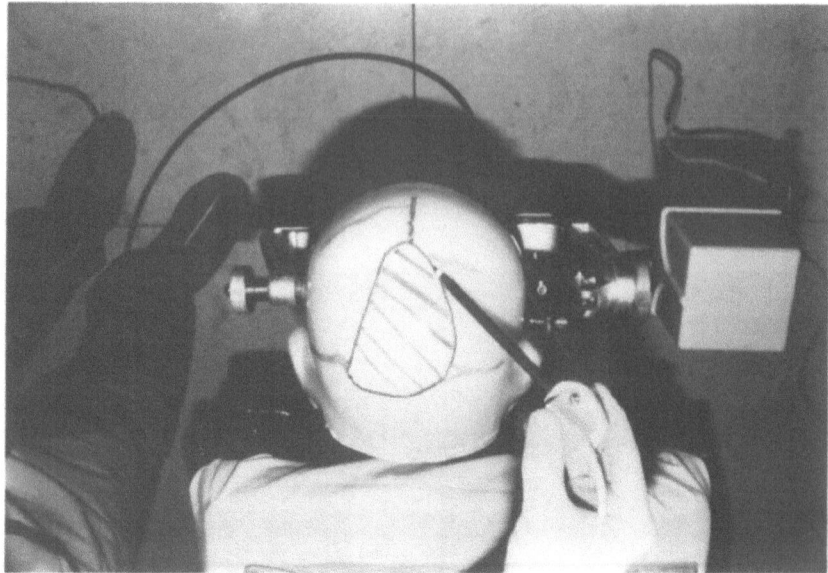

a   b

Fig. 9. (a) By using the dummy head as a preoperative planning tool, the patient's magnetic resonance angiogram from an occipital approach can be mapped. (b) Various color marking pens are substituted as instruments and attached to the receiving antenna. Location of the major draining sinus and contributing cortical venous tributaries allows planning of an optimal bone flap to gain access to the torcular neoplasm

lenscope is displayed simultaneously on the monitor. Figure 9 shows the coronal importation of a magnetic resonance angiogram image with unusual vascular hemangioma of the torcula. Surgical planning including scalp flap, bony window margins and location of transverse sinus, sagittal sinus and draining cortical venous tributaries was performable on the dummy head model as shown.

Discussion

Our laboratory trials for accuracy of the magnetic field guided intracranial positioning system suggest sufficient precision of measurement for most neurosurgical applications including routine preoperative planning, incisional planning, anthropometric documentation and cranial reconstruction, cerebral and lateral ventricular lesions. The accuracy is probably not acceptable for brainstem lesions. The composite accuracy using our dummy model with multiple implanted spheres and burr holes shows a deterioration in accuracy to 5 mm. This is consistent with the additive effect of our curve-fitting translation of the coordinate system of the radiographic images sphere and the small inaccuracy of the CT scanner, filming, and video capture technique. When the receiving antenna

is attached to the endoscope, the accuracy is consistently sufficient to bring into the view angle the target, allowing direct contact under visual guidance. Thus, it is reasonable that magnetic field guidance will significantly enhance surgical localization and orientation compared to the unaided endoscope.

The greatest limitation for extensive endoscopic surgery is bleeding and adequate methods of hemostasis. The avoidance of major vessels due to the coupled intracranial positioning system is a reasonable first line of strategy.

Three impediments have been recognized to date in our present system configuration. The receiving antenna cannot yet be made sufficiently physically small to be mounted at the tip of the steerable fiberscope. Therefore, the receiving antenna must remain external to the cranium. An intracranial path must always therefore be a direct trajectory. Ventricles and cyst cavity may change shape and volume when entered. This problem is less serious compared to open craniotomy, but nevertheless renders preoperative CT and MRI maps less accurate. Finally, maintenance of a precise and fixed relationship between the transmitter antenna and the cranium is difficult with observed deterioration of accuracy over the course of a procedure. While several strategies for detection of this drift have been proposed, including reregistration with fiducial markers, each is limited due to the small area of exposure of a sterile surgical field.

Despite these impediments we believe magnetic field guidance as a form of "intracranial radar" coupled with the neuroendoscope opens the door for much more effective or extensive endoscopic dissection or debulkment.

References

1. Hassenbusch SJ, et al (1991) Brain tumor resection aided with markers placed using stereotaxis guided by magnetic resonance imaging and computed tomography. Neurosurgery 28: 801–806

2. Heilman CB, Cohen AR (1991) Endoscopic ventricular fenestration using a "saline torch." J Neurosurg 74: 224–229

3. Kato A, et al (1991) A frameless, armless navigational system for computer-assisted neurosurgery. J Neurosurg 74: 845–849

4. Manwaring KH (1992) Neuroendoscopy Vol. 1. Chapter—Endoscopic ventricular fenestration. Liebert, New York

5. Manwaring KH, Beals SP (1989) Bloodless dissection in pediatric craniotomies and craniofacial surgeries (Abstract). American Association of Neurological Surgeons; Section on Pediatric Neurological Surgery. Washington, DC

6. Watanabe E, et al (1991) Open surgery assisted by the neuronavigator, a stereotactic, articulated, sensitive arm. Neurosurgery 28: 792–800

Correspondence: Kim H. Manwaring, M.D., Section of Pediatric Neurosurgery, Phoenix Children's Hospital, 909 East Brill Street, Phoenix, Arizona 85006, U.S.A.

Acta Neurochir (1994) [Suppl] 61: 40–42

New Kinds of Microneuroprotectors for Microsurgery and Endoscopy of Cerebellopontine Angle Neurovascular Decompression

A.A. Khodnevich and **V.B. Karakhan**

Department of Neurology and Neurosurgery, Moscow Medical Stomatological Institute, Moscow, Russia

Summary

New microsurgical devices for neurovascular decompression—microneuroprotectors (MNP)—are described. Four constructive kinds of MNP have been developed according to topographic peculiarities of pathological neurovascular contacts. The hydrodynamic and biological testing of MNP has been concluded. The methods of microsurgical and endoscopic techniques of MNP insertion on the cranial nerves or posterior fossa vessels are reported.

Keywords: Cerebellopontine angle; cranial nerves; endoscopy; neurovascular decompression; trigeminal neuralgia; surgery.

Introduction

Relapse rate and noneffectiveness of microvascular decompression of the V, VII, VIII, IX cranial nerves in an entry-exit zone of the brain stem vary from 8 to 12–36%[1–4,8,9]. This may be caused by: 1) slipped protect prothesis from the zone of neurovascular contact[4]; 2) insufficient insulation quality of the protecting prothesis under the compressive arterial pulsation[4,6]; 3) the recanalization of sectioned veins with the renewal of nerve compression root[4]; 4) development of connective tissue adhesions; 5) missing of plural and combined decompression because of one-sided review of the nerve root or a violation of natural anatomic-topographic relationships in displacing the cerebellum, which is the disadvantage of the direct approach[3,4]. Therefore protection of the nerve root along the length of its intracranial portion, increasing of the suppressive quality of the protecting prothesis and development of new surgical approaches are necessary.

Devices and Methods

A series of different kinds of microneuroprotectors (MNP) has been developed (Fig. 1). MNP has the form of a cylindrical cuff cut lengthwise, there is a bulge opposite the slit. While squeezing the bulge with tweezer blades the MNP opens; it enables the user to apply the MNP and to embrace a nerve or a vessel (Fig. 2).

Four types of MNP are offered (Fig. 1).

1. Latex or solid silicon MNP with elliptical cuff contours embraces an artery or nerve in the case of compression of the nerve root by a vein or artery which should not be sectioned. This MNP redistributes static pressure on the vessel over a bigger area, and also protects the nerve in case of recanalization of a coagulated and sectioned vein (Fig. 1a).
2. Porous silicon MNP embraces a nerve or an artery in a case of typical compressive arterial pulsation of the loop of a cerebellar artery (Fig. 1b).
3. MNP with a double isolated wall (Fig. 1c), which has very high protective quality, embraces a nerve or a vessel in the case of compression of a nerve root by an artery.
4. Like 3 but with the apron defending the pons near the trigeminal root entry zone from arterial loop contact (Fig. 1d).

The silicon MNP is formed from a module, which has the shape of a truncated cone. The diameters of the large and the small bases are 6.0 mm and 1.0 mm, the length is 20 mm.

We have performed a comparative evaluation of the protective quality of these prothesis under the conditions of stand hydrodynamic design. The standard model consists of: 1) a model of an arterial vessel—of curved latex tube with a diameter 1.0 mm—through which a liquid was pumped under the changeable pressure of 120/80 mm Hg, with a frequency of 70 Hz; 2) a model of a nerve in the form of a tube with a diameter 3.0 mm with a hole on the surface where a sensitive membrane is put. The tube is hermetically connected with a data unit of pressure. While pressing the membrane the change of gas pressure in the tube is registered by the data unit, transformed into an electric signal and depicted graphically by an amplifier. The models of "vessel" and "nerve" were put perpendicular and touch each other. The construction was plunged into water under pressure of 20 mm water column which is correspondent to CSF pressure in the cisterns of the cerebellopontine angle.

For assessment of tissue reaction MNP have been implanted on the tibial nerve of white rats (from 1 week to 4 months). Subsequent neuropathological investigations have been performed.

Results

A comparative hydrodynamic testing of various MNP is presented in Fig. 3. The result is that application of

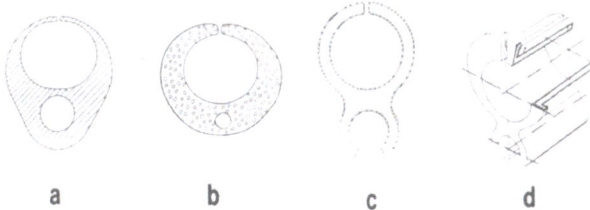

Fig. 1. (a–d) Different kinds of microneuroprotectors

Fig. 2(a, b). Constructive peculiarity providing for ease of application of MNP by microforceps

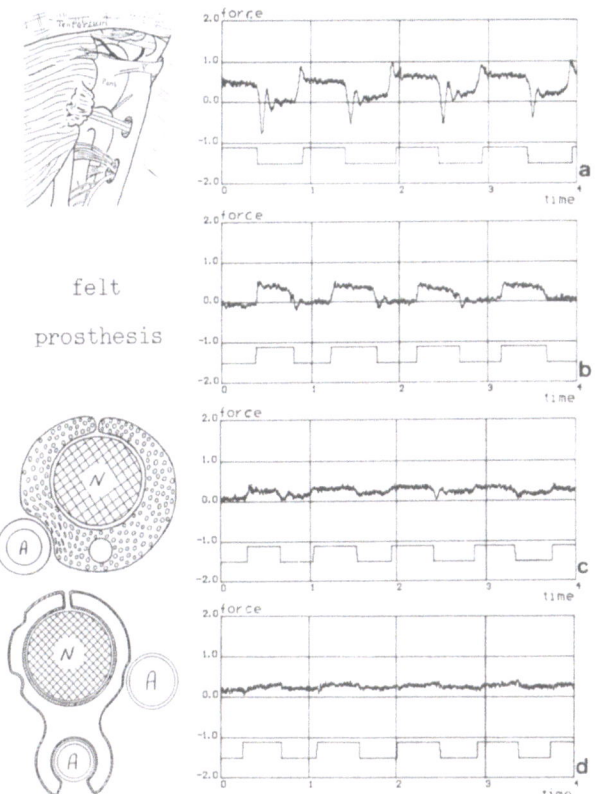

Fig. 3. Comparison of hydrodynamic testing of various MNP. (a) Transfer of pulsatile influence of "a vessel" upon "a nerve" without a protective prosthesis. (b) Partial lowering of pulsatile influence using felt piece. (c, d) Nearly complete absorbtion of a pulsatile influence of "a vessel" upon "a nerve" by porous silicon MNP and MNP with a double isolated wall

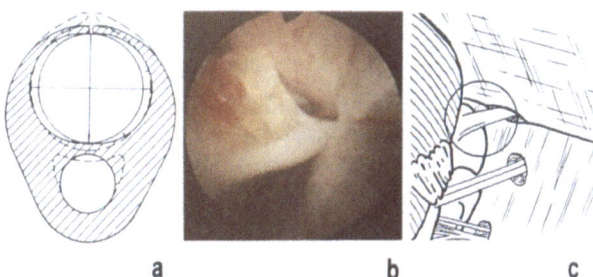

Fig. 4. Construction properties of solid MNP for installation on an artery: (a) Change of blades and bridge position during concentric enlargement of an artery with minimal shift of outer contours of the device. (b) MNP embracing superior cerebellar artery loop separating it from the trigeminal root; – endofiberoscopic picture; (c) stereotopography of endoscopic view

porous silicon MNP and MNP with a double isolated wall installed on a nerve provides a 6–10 fold decrease of pulsative transfer. Moreover the last kind of MNP can prevent a tangential influence of an arterial loop on a nerve.

The essentials of solid MNP for installation on the artery are shown in Fig. 4.

The histological data show a thin collagen membrane around the MNP without axonal destruction or inflammatory reactions.

The porous silicon MNP has been used in 22 patients with various forms of neurovascular compression in the cerebellopontine angle. There were no complications.

Discussion

Various types of plastic and biological protective devices are used to preserve the nerve-vessel insulation after neurovascular decompression[1–5, 8]. But application of these devices cannot exclude their migration and perifocal adhesional activity. X-Ray contrast sponges provide the diagnosing of such an event but not the prevention of the prosthesis shift[7]. The usage of Sundt's metallic clips[5] does not ensure the high protection and can damage the nerve root by rigid clip borders.

The advantages of our original microneuroprotectors are:

1. Concentric protection of a nerve against multiple compression factors and connective tissue adhesions along the whole length of the root.
2. Prevention of postoperative MNP migration.
3. High protective qualities.

4. Atraumatic intracranial introduction and installation on a nerve or artery by very elastic substantion and distant controlled positioning of MNP.
5. X-Ray, CT, MRI checking of MNP position.
6. Minimal biological reaction to MNP implantation.
7. Possibility of MNP installation by endoscopic instruments.

The endoscopic method ensures minimal displacement of the cerebellum, avoids alteration of normal anatomical relations of neurovascular contacts, preventing multiple neural compression and also excessive stretching of the facial and vestibulocochlear nerves.

References

1. Apfelbaum RI (1982) Microvascular decompression for tic doloureux: results. In: Brackmann PE (ed) Neurological surgery of the ear and skull base. New York, pp 175–180
2. Breeze R, Ignelzi RJ (1982) Microvascular decompression for trigeminal neuralgia. Results with special reference to the late recurrence rate. J Neurosurg 57: 487–490
3. Jannetta PJ (1977) Treatment of trigeminal neuralgia by suboccipital and transtentorial cranial operations. Clin Neurosurg 24: 538–549
4. Jannetta PJ, Bissonette OJ (1984) Management of failed patient with trigeminal neuralgia. Clin Neurosurg 32: 334–347
5. Laws ER, Kelly OJ, Sundt TM (1986). Clip-graft in microvascular decompression of the posterior fossa. J Neurosurg 64: 679–681
6. Nagahiro S, Takada A, Matsukado Y et al (1991) Microvascular decompression for hemifacial spasm. Patterns of vascular compression in unsuccessfully operated patients. J Neurosurg 75: 388–392
7. Nagata K, Sasaki T, Basugi N (1986) Radiopaque synthetic sponge as a prosthesis for microvascular decompression. Technical note. J Neurosurg 65: 564–565
8. Szapiro J, Sindou M, Szapiro J (1985) Prognostic factors in microvascular decompression for trigeminal neuralgia. Neurosurgery 17: 920–929
9. Zorman G, Wilson CB (1984) Outcome following microsurgical vascular decompression or partial sensory rhizotomy in 125 cases of trigeminal neuralgia. Neurology 34: 1362–1365

Correspondence: V.B. Karakhan, M.D., Department of Neurology and Neurosurgery, Moscow Medical Stomatological Institute, Delegatskaya str. 20/1, 103473 Moscow, Russia.

Acta Neurochir (1994) [Suppl] 61: 43–45

A Computer Assisted Toolholder to Guide Surgery in Stereotactic Space

C. Giorgi, F. Pluchino, M. Luzzara, E. Ongania, and **D.S. Casolino**

Department of Neurosurgery, Istituto Nazionale Neurologico "C.Besta", Milano, Italy

Summary

A computer assisted toolholder, integrated with an anatomical graphic 3-D rendering programme, is presented. Stereotactic neuro-anatomical images are acquired, and the same reference system is employed to represent the position of the toolholder on the monitor. The surgeon can check the orientation of different approach trajectories, moving the toolholder in a situation of virtual reality. Angular values expressed by high precision encoders on the five joints of the toolholder modify "on line" the representation of the configuration of the toolholder within the three dimensional representation of the patient's anatomy.

Keywords: Stereotaxy; computer assisted toolholder.

Introduction

Stereotaxy, the art of reaching targets within the brain based on properly acquired neuroradiological examination, was developed and has been practiced so far with mechanical aiming devices. Arc-centered or interlocked-angle instruments have all proved satisfactory for the task for which they were designed. The introduction of guided open neurosurgery since highly detailed neuroanatomical images in stereotactic conditions have become available, now requires improvement in neurosurgical instrumentation to take full advantage of the available information.

Currently, the use of traditional stereotactic frames, designed to guide a probe to a target point within the stereotactic space, provides tangible improvements in the ability to reach deep seated lesions and remove them through narrow incisions[2,3]. Nevertheless, the limitations of an instrument that is cumbersome to handle for entering different settings during the process of removing a volume larger than a few millimeters in diameter, soon becomes evident when put into practice.

In order to alleviate the burden of setting angles and depths during neurosurgical procedures, with the attendant risks of errors and contamination of read-ing millimetric scales under surgical drapes, we have developed a computer controlled toolholder with five free joints.

We think that this instrument can be developed into an intelligent tissue retractor or a guide for surgical tools (including echographic probes) which will include all the information provided by a conventional stereotactic apparatus.

Materials and Methods

The prototype toolholder described here consists of an articulated arm with five high precision encoders (1/40th of a degree) on the joints, each of which has a fail-safe electromagnetic brake. The arm is able to support 300 grams at the tip when extended. Each encoder sends a stream of readings to the graphic processor, through a multiplexer (Fig. 1).

The arm provides real time knowledge of position and orientation of instruments mounted on the end effector. To gather this information, the angular position of each joint and the dimensions of the links, have to be known. Joint angles are detected by differential incremental encoders, in steps of 1/40th of a degree. The lengths of the links were determined, so as to be transferable onto the kinematic model instead of those indicated in the drawings. Based on the direct kinematics, the theoretical error of various positions of the end effector was estimated by introducing the maximum possible error (1/40th of a degree) at each of the five joints. The theoretical error reads between 0.2 and 0.4 millimeters, this last figure being for positions of the arm in full extension.

To evaluate the accuracy of the system in practice, the arm is mounted on an X, Y, Z positioner, with a precision greater then one hundredth of a millimeter. The tip of the end effector is located on a starting point that is selected as the origin of the calibration, with the error set to zero. The accuracy of the system is determined for a cluster of points over a three dimensional grid of 2.5 mm mesh. This process shows that the error ranged between 0.5 and 0.8 mm, and that the module of the error increases with increasing distance of the point from the origin. This provides the possibility of correcting position errors via software. In fact, knowing the error at each point of the grid, it is possible to estimate a positioning error for each point of the working areas, and use that to correct the readings expressed by the direct kinematics.

Once the arm has been calibrated, its position can be calculated, with respect to the stereotactic frame by reading the positions of the origin and those of three points along x, y and z axes with the tip of

Fig. 1. The toolholder, resting on a fast-calibration bench, which allows cheching of readings from the encoders when the links rest on calibrated posts

the end effector. This is accomplished with a "phantom" localizer of a stereotactic frame, and the result provides a common reference system between the frame and the arm. Using a graphic computer, each segment of the articulated arm and the stereotactic frame is modelled three dimensionally, according to the drawings. CAD modelling techniques were employed.

The operative programme accepts CT, MRI and angiographic images of the patient under scrutiny, acquired under stereotactic conditions, and enables the surgeon to trace the contours of lesions and surrounding intact cerebral structures. The programme reconstructs the volumes of the structures outlined within the coordinate system of the stereotactic frame (Figs. 2 and 3).

The output of the five encoders of the arm enter the surgical graphic programme via a serial port and, in agreement with the kinematics, represents the arm segments in the configuration that the instrument assumes when moved by the surgeon. There are several ways to help the surgeon perceive the third dimension while

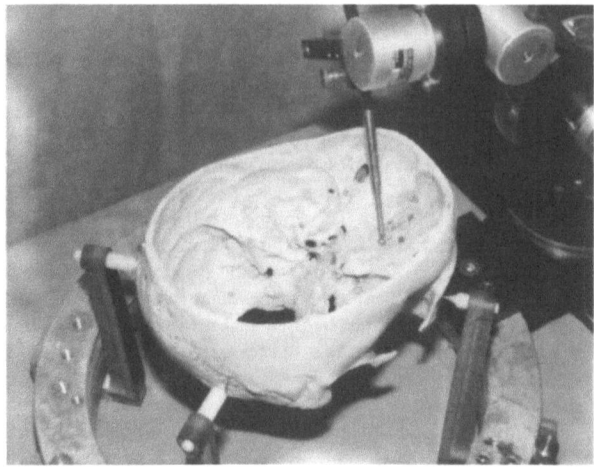

Fig. 2. A metal target glued on a dry skull, touched with a probe mounted on the toolholder. Skull has been previously scanned, with a CT localizer mounted on the stereotactic frame

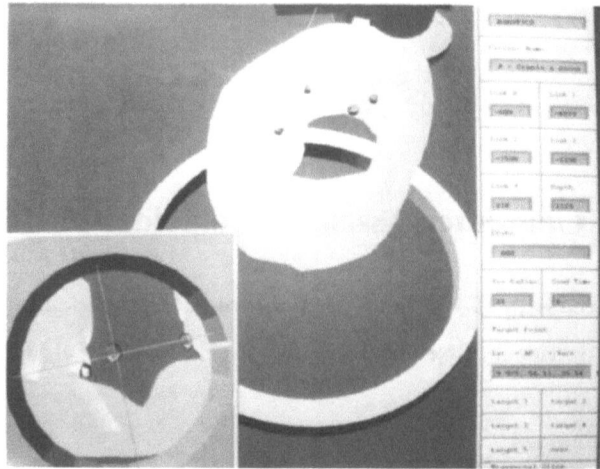

Fig. 3. 3-D reconstruction of the skull and metal targets, obtained by CT, stereotactically acquired. Probe position and toolholder orientation are calculated from the values of the encoders. Perfect correspondence with actual probe position is evident in Fig. 2. LL quadrant insert shows "surgeon's eye view"

localizing the target, at the same time avoiding "eloquent" parts of the brain. A "surgeons's eye-view" or the use of stereoscopic goggles are among the most efficient of these resources.

Discussion

Preliminary clinical testing of the instrument satisfies our expectations: this tool can be substituted for the stereotactic arc, whose function is limited to the identification of a trajectory to a target point, and has the added function of serving as a holder of surgical instruments. Its use, mastered in minutes by every neurosurgeon who has tried it, eliminates possible errors that might be introduced by setting wrong numbers on a traditional stereotactic arc. The precision of the present prototype is satisfactory, because the error is confined within the size of a voxel of images used for stereotactic localization. This result can only be obtained if images are referred to a head fixation device that is rigidly linked to the toolholder. At present, only a stereotactic frame can offer the required rigidity. Fail-safe brakes, dependable encoders, and backup power supplies contribute to the safety of the computer-assisted toolholder. In our opinion, other localizing devices based on ultrasound, laser light, magnetism or stereoscopic acquisition of light arrays, lack the necessary simplicity to fit the scenario of a microsurgical operating field, without providing significantly greater accuracy[6,7].

This impression is shared by other investigators,

who have described mechanical localizers that have partially anticipated the solution offered by our instrument[1,4,5,8]. Since these often lack stereotactic frames and have graphic rendering software that is too simplified, we have decided to continue with our design, working on improving its precision and on refining the graphic description of anatomical information. Future work will be concentrated on the implementation of an echographic probe mounted on the tool, or positioned with its aid, in order to obtain on line stereotactic echographic data to monitor anatomical changes during surgery.

References

1. Drake JM, Joy M, Goldenberg A, Kriendler D (1991) Computer and robot assisted resection of thalamic astrocytomas in children. Neurosurgery 29: 27–33
2. Giorgi C, Ongania E, Casolino SD, Riva D, Cella G, Franzini A, Broggi G (1991) Deep seated cerebral lesion removal, guided by volumetric rendering of morphological data, stereotactically acquired. Clinical results and technical considerations. Acta Neurochir (Wien) [Suppl] 52: 19–21
3. Kelly PJ, Alker GJ, Goerss S (1982) Computer-assisted stereotactic laser microsurgery for the treatment of intracranial neoplasms. Neurosurgery 10: 324–331
4. Kwoh YS, Hou J, Jonckheere EA, Hayati S (1988) A robot with improved absolute positioning accuracy for CT guided stereotactic brain surgery. IEEE Trans Biomed Eng 35: 153–160
5. Reinhardt HF, Landolt H (1989) CT-guided real time stereotaxy. Acta Neurochir (Wien) [Suppl] 46: 107–108
6. Roberts DW, Strohbehn JW, Hatch JF, Murray W, Kettenberger H (1986) A frameless stereotactic integration of computerized tomographic imaging and the operating microscope. J Neurosurg 52: 21–27
7. Tsai RY (1987) A versatile camera calibration technique for high accuracy 3-D machine vision metrology using off the shelf TV cameras and lenses. IEEE J Robot Auto 4: 323–344
8. Watanabe E, Watanabe T, Manaka S, Mayanagi Y, Takakura K (1897) Three dimensional digitizer (neuronavigator): new equipment for computed tomography—guided neurosurgery. Surg Neurol 27: 543–547

Correspondence: C. Giorgi, M.D., Department of Neurosurgery, Istituto Nazionale Neurologico "C.Besta", Via Celoria 11, I-20133 Milano, Italy.

Acta Neurochir (1994) [Suppl] 61: 46–48
© Springer-Verlag 1994

Ultrasound Guided Endoscopic Neurosurgery—New Surgical Instrument and Technique

K. Yamakawa, T. Kondo, M. Yoshioka, and **K. Takakura**[1]

Department of Neurosurgery, National Medical Center and [1]University of Tokyo Hospital, Tokyo, Japan

Summary

For minimal invasive endoscopic neurosurgery, a mini-caliber endoscope equipped with a useful working channel and its guidance system is basically essential. We developed the device for ultrasound guided endoscopic neurosurgery.

A minicaliber fiberscope with an outer diameter of 3.4 mm equipped with 2.5 mm of working channel for forceps, suction, and coagulation probe was developed. A 5 MHz phase arrayed 12 mm ultrasound probe was used as guidance system. A fixation device for the mini-caliber fiberscope and ultrasound probe was also constructed. The surgical procedure is very simple, and can be performed through a 2 cm burr hole. Real-time and a directly visualized image might provide more safety and reliability during brain surgery.

Keywords: Neuroendoscopy; ultrasound.

Introduction

Since the first decades of this century, many practical uses of endoscopes for ventriculoscopy[3,7,9,11,13,15] and myeloscopy[4,10,15] have been developed in the field of neurosurgery. Recently the development of fiberoptic technology has improved endoscopes in resolution and miniaturization.

Endoscopy in combination with either CT[4,14,17], MRI[5] stereotaxy, or ultrasound[1,2] as a guiding device for the endoscope has made it possible to operate on lesions located not only in the ventricular system but also in the hemispheres.

We have used superfine (ultrathin) fiberscopes for ventriculoscopy, myeloscopy, and endovasculoscopy. Its usefulness has been reported elsewhere[15]. The sizes of the instruments provide minimally invasive surgery for brain and neural tissue. However, the visual field is restricted so that anatomical orientation can not be accurately obtained. We developed a device for ultrasound guided endoscopy (US-Guided Endoscopy) using a minicaliber fiberscope equipped with an appropriately sized working channel.

The intention of the present study is to describe the surgical apparatus of ultrasound guided endoscopy and present the advantages of this technique and the limitations of its clinical applications.

Instruments and Operative Technique

Surgical Devices (Fig. 1)

For neuroendoscopy, a rigid fiberscope covered with stainless steel with an outer diameter of 3.4 mm, effective length of 8 cm, and equipped with 2.5 mm working channel was developed (Medical Science Co., Ltd). The outer tube with an guiding mandrin (outer diameter of 4 mm) is used to prevent unnecessary damage of brain tissue by the tip of the endoscope. Microsurgical instruments such as insulated forceps, coagulation probe, and suction tube were also constructed. A 5 MHz phase arrayed electric sector scanner with a diameter of 12 mm is used for guiding the outer tube to the target lesion (ALOKA Co., Ltd). The fixation device for the ultrasound probe made of stainless steel is attached to the skull (Mizuho Medical Co., Ltd). This fixation device has 2 joints, and the ultrasound probe can be fitted to the surface of the brain.

Surgical Procedure

The patient's head is positioned so that the burr hole can be placed in a horizontal plane. Surgery can be performed under local or general anaesthesia, depending on the patient's co-operation and the kind of surgical procedure. The site of the burr hole is determined by CT examination taking into consideration the distance from the brain surface to the target lesion, direction of surgical approach, and the importance of minimal damage to normal brain tissue. After burr hole trepanation the fixation device for the ultrasound probe is attached to the skull. The outer tube is inserted into the brain along the imaginary target line on the ultrasound monitor. The tip of the outer tube reaches the lesion, the guiding mandrin is withdrawn and the neuroendoscope is inserted.

Clinical Application

Under direct visualization, surgical manipulations such as incision, biopsy, evacuation, irrigation, drain-

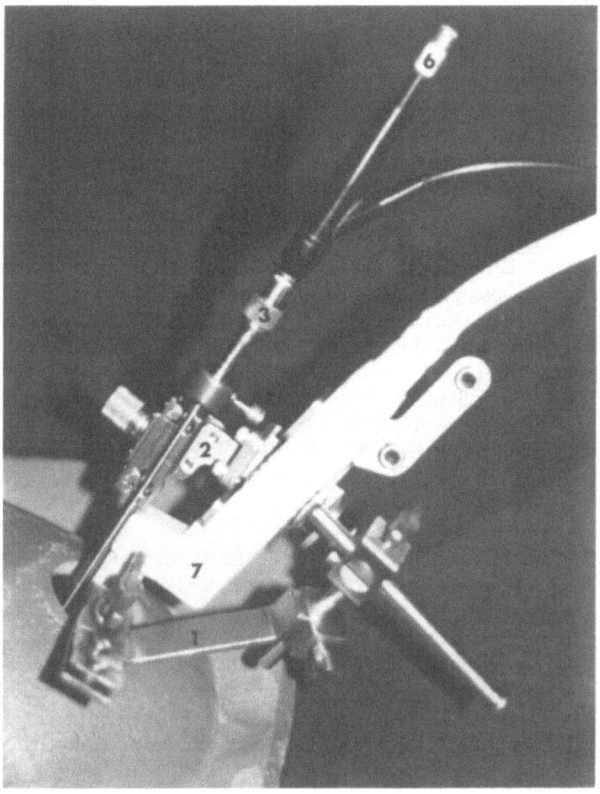 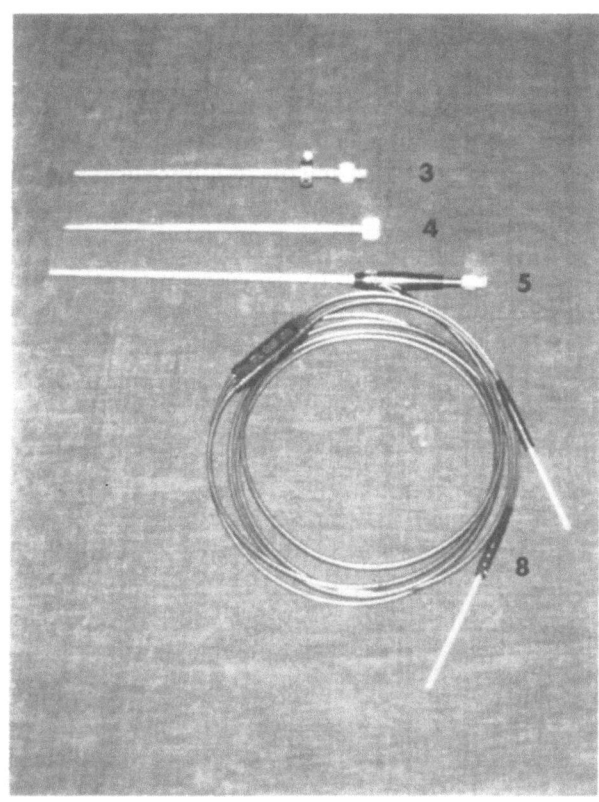

a   b

Fig. 1. (a, b) Device of ultrasound guided endoscopy. *1* fixation device, *2* puncture adapter, *3* outer tube, *4* guiding mandrin, *5* neuro-endoscope, *6* working channel, *7* ultrasound probe, *8* image and light guide

Table 1. *Surgical Manipulations in US-Guided Endoscopy*

Incision
Biopsy
Removal
Evacuation
Irrigation
Coagulation
Drainage
Ventriculostomy
Laser Surgery

Table 2. *Surgical Application of US-Guided Endoscopy*

Intracerebral haematoma
Intraventricular haematoma
Brain tumour
Hydrocephalus
Brain abscess
Ventriculitis
Parasite
Cavernous angioma
. . . . . etc.

age, coagulation, and laser surgery can be safely performed (Table 1). Possible indications for US-guided endoscopy are: intracranial haematoma, cystic or small brain tumours, hydrocephalus, and brain abscess. (Table 2).

Discussion

In the last decade, different ultrathin fiberscopes have been clinically used for observations of small and narrow structures of the human body such as coronary

artery[8], urinary tract[16] tympanic cavity[6], and for arthroscopy[12]. In neurosurgery, we started using ultrathin fiberscopes for ventriculoscopy, myeloscopy and endovasculoscopy in 1987[15]. Thin endoscopes certainly minimize traumatization of brain and neural tissue, however, the visible field is restricted, so it is difficult to get precise anatomical orientation.

In our study, ultrasound is considered to be helpful for guiding the endoscope to the target region. Furthermore, ultrasound provides useful information during surgery with regard to the structural changes of the lesion and its surroundings. At present, we only use a 5 MHz phase arrayed ultrasound probe. This probe is small and provides a clear image of deep seated lesions, however, in small subcortical lesions, the image is unfocused and the resolution is poor. The use of a probe with a higher frequency of 7.5 MHz or 10 MHz excellently demonstrates the image near the probe and is thought to be necessary for routine investigation.

The advantages of US-guided endoscopy are real time imaging, direct visualization, minimal invasion, and a safe, simple procedure. In comparison with conventional CT-guided or ultrasound-guided stereotaxy, more aggressive surgical procedures such as irrigation, coagulation, removal, and laser surgery can be performed. At present, we use the rigid fiberscope without angulation systems. Despite of some restriction in the visual field, the minicaliber flexible fiberscope with angulation system is thought to be useful for intraventricular endoscopic observation and surgery.

References

1. Auer LM, Holzer P, Ascher PW (1988) Endoscopic neurosurgery. Acta Neurochir (Wien) 90: 1–14
2. Auer LM, Deinsberger W, Kurt Niedarkorn K, Gell G, Kleinert R, Schneider G, Holzer P, Bone G, Mokry M, Korner E, Kleinert G, Hanusch S (1989) Endoscopic surgery versus medical treatment for spontaneous intracerebral hematoma: a randomized study. J Neurosurg 70: 530–535
3. Griffith HB (1987) Endoneurosurgery: endoscopic intracranial surgery, In: Wenker H *et al* (eds) Advances in neurosurgery, Vol 14. Springer, Berlin Heidelberg New York, pp 2–24
4. Hellwig D, Bauer BL (1992) Minimally invasive neurosurgery by means of ultrathin endoscopes. Acta Neurochir (Wien) [Suppl] 54: 63–68
5. Hor F, Desgeorges M, Rosseau GL (1992) Tumor resection by stereotactic laser endoscopy. Acta Neurochir (Wien) [Suppl] 54: 77–82
6. Kimura H, Yamaguchi H, Cheng S, Okudaira T, Kawano A, Iizuka N, Imakirei M, Funasaka S (1989) Direct observation of the tympanic cavity by the superfine fiberscope. J Otolaryngol Jpn 92: 233–238
7. Kleinhaus S, German R, Sheran M, Shapiro K, Boley SJ (1982) A role for endoscopy in the placement of ventriculoperitoneal shunts. Surg Neurol 18: 179–180
8. Mizuno K, Arai T, Satomura K, Shibuya T, Arakawa K, Okamoto Y, Miyamoto A, Kurita A, Kikuchi M, Utsumi A, Takeuchi K (1989) New percutaneous transluminal coronary angioscope. J Am Coll Cardiol 13: 363–368
9. Oka K, Ohta T, Kibe M, Tomonaga M (1990) A new neurosurgical ventriculoscope: technical note. Neuro Med Chir 30: 77–79
10. Ooi Y, Satoh Y, Hirose K, Mikanagi K, Morisaki N (1977) Myeloscopy. Int Orthop 1: 107–111 6: 881–894
11. Putnam TJ (1934) Treatment of hydrocephalus by endoscopic coagulation of the choroid plexus. Description of a new instrument and preliminary report of results. N Engl J Med 210: 1373–1376
12. Sawai K, Ishibashi K, Asada K, Hamada K, Naohara H, Yamanaka K, Jibiki M, Kobayashi K (1990) Development and application of fine needle fiber-arthroscope to the T.M.J. Diagnostic use of the system. JJOMS 36: 423–433, 1990
13. Scarff JE (1952) Non-obstructive hydrocephalus. Treatment by endoscopic cauterization of the choroid plexus. J Neurosurg 9: 164–176
14. Shelden CH, Jacques DB (1982) Neurologic endoscopy. In: Schmidek HH, Sweet WH (eds) Operative neurosurgical techniques, indications, methods and results, Vol 1. Grune and Stratton, New York, pp 419–431
15. Yamakawa K, Kondo T, Yoshioka M, Takakura K (1992) Application of superfine fiberscope for endovasculoscopy, ventriculoscopy, and myeloscopy. Acta Neurochir (Wien) [Suppl] 54: 47–52
16. Yoshida K, Nishimura T, Tsuboi N, Hasegawa J, Kawamura N, Chorazy ZJ, Akimoto M (1991) Clinical application of video image flexible ureteronephroscope for diagnosis of upper urinary tract disorders. J Urol 146: 809–812
17. Zamorano L, Chavantes C, Dujovny M, Malik G, Ausman J (1992) Stereotactic endoscopic interventions in cystic and intraventricular lesions. Acta Neurochir (Wien) [Suppl] 54: 69–76

Correspondence: Kenta Yamakawa, M.D., Department of Neurosurgery, National Medical Center, 1-21-1, Toyamacho, Shinjuku-ku, Tokyo 162, Japan.

Acta Neurochir (1994) [Suppl] 61: 49–53

A Multipurpose Cerebral Endoscope and Reflections on Technique and Instrumentation in Endoscopic Neurosurgery

J. Caemaert, J. Abdullah, and **L. Calliauw**

Department of Neurosurgery, Hospital University Ghent, Belgium

Summary

We discuss our experiences concerning our cerebral endoscope with reflections on various techniques used since 1986. During this time we have had experience with four prototypes.

This minimal invasive procedure has been successful to a certain extent both in paediatric and adult patients, stereotactically and by freehand method or both.

Further modification for flexibility and manipulation of the optic element is under development.

Keywords: Cerebral endoscopy; techniques; instrumentation.

Introduction

Renewed interest in endoscopic neurosurgery is mainly due to the recent developments in neuro-imaging. An important variety of lesions is situated in the para- or intraventricular regions. Other lesions are cystic. The excellent visualization of these anomalies by CT and MRI scan enables one to make a precise approach and plan strategies.

Only recently highly suitable and specific (cerebral) instrumentation permitting us not only to inspect but also to operate is being developed[1, 3, 4, 5, 6].

In 1986 we designed a multipurpose endoscope which was realised by Richard Wolf, Belgium, in association with R.W. Knittlingen, Germany (Fig. 1). Since then we have used four prototypes until April 1992.

Methods

The instrument consists of a long rigid shaft (30,5 cm) and a set of 3 stop-cocks at the proximal end of three of the four inner channels (Fig. 2). One channel contains the optic element (I.D. 3 mm), the second one permits the introduction of different "working instruments" (I.D. 2,4 mm) (Fig. 3). Two smaller channels

(I.D. 1,67 mm) permit rinsing with saline and suction. To one of them an infusion set is coupled with a 5 liter bag of normal saline (36° C), and to the other one an open infusion tube is connected to collect the outcoming fluid into a collecting receptacle. This latter should be installed at the zero pressure level to prevent a siphoning effect with acute collapse of the ventricles. This phenomenon obliterates any view and entails the risk of tearing of cortical drainage veins and the development of a subdural haematoma. If this collapse occurs, the outlet cock has to be closed and rinsing abundantly with saline solution is necessary. It is also important to mention that even a partial collapse or even diminishing of the ventricular dilatation may considerably alter the position of the periventricular target areas with complete loss of orientation. It is necessary to have the possibility of rinsing since even a very small amount of blood causes blurred vision that prevents working and recording of images. It should be stressed that such small bleedings occur nearly always, even with perfect technique. It comes for example from small vessels in the ventricular wall, there where it is perforated by the shaft and thus out of reach.

During the whole procedure intracranial pressure can be monitored by means of an epidural or intra-parenchymatous pressure sensor. Rinsing should not be started before reaching the ventricle or cystic lesion otherwise severe oedema in the white matter may be caused.

The length of the shaft is necessary to permit the introduction through the carrier of the stereotactic arc since the video camera and the light source cable are the largest "blocking" elements. The cross section of the shaft should be round to permit turning in all directions around its own axis. This is important be-

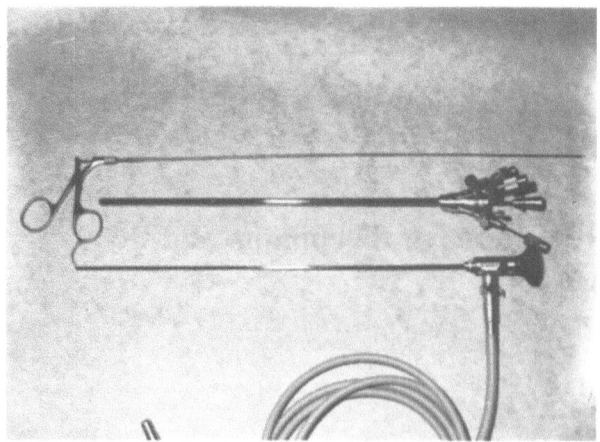

Fig. 1. Multipurpose endoscope, developed by Richard Wolf, Belgium, in association with R.K. Knittlingen, Germany

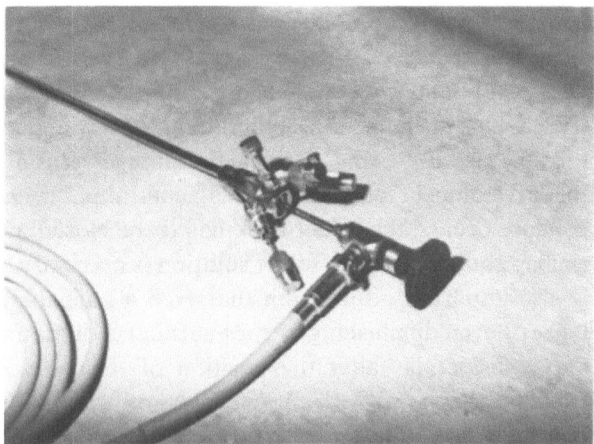

Fig. 2. Details of the multipurpose endoscope

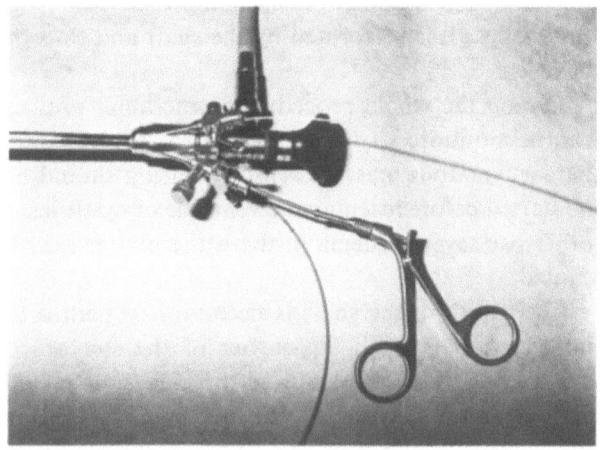

Fig. 3. Introduction of "endoscopic instruments"

cause, by turning, the field of action is considerably enlarged thanks to the 10° angled optic arrangements and the precurved tips of some flexible instruments. In the actual (fourth) prototype the outer diameter is 6 mm.

The question of whether a cerebral endoscope should be rigid or flexible is irrelevant. The real question is when one should use a rigid or a flexible one. The advantage of the usual flexible endoscopes[3] is the small outer diameter (3,1 mm). The drawback is the lack of several channels for simultaneous rinsing and working. This is extremely important since even a small amount of bleeding prevents any action. A flexible optic element introduced through the working channel of our rigid endoscope is in preparation. This will enable "looking round the corner" after a linear penetration of for example the septum pellucidum.

The instrument can be used stereotactically as well as for free hand interventions. It is introduced through a single 8 cm-long guide, fixed either to the stop-and-guide-carrier of any stereotactic frame or to a special support mounted on the table and serving as an optional aid under free hand conditions. This support however never replaces the hands of a trained assistant who may adjust the position of the endoscope by very delicate movements on request of the surgeon.

The classical stop and guide are replaced by a single 8 cm-long cylindrical guide with an asymmetrically placed lumen of 6 mm. The reason for this is that the working channel lies asymmetrically in the shaft opposite the optic channel and each of them can be introduced according to the central stereotactic axis leading to the target. Because of the asymmetry a separate stop and guide would cause immediate blockage of the shaft when the two lumina are not perfectly coaxial.

Each channel can be filled with a mandrin during the introduction through cortex and white matter. The optical element, however, is most often in place from the beginning to permit immediate visualisation of the penetration of ependyma or cyst walls. In many cases a stereotactic approach to the target is indicated but once inside the ventricle or cystic structure, the guide may be detached from the carrier on the arc. After this one can move freely under direct visual control. Obviously extensive "rowing" in the white matter should be avoided but gentle careful movements permit visualization of a large field of action. Postoperative MRI shows very little damage along the shaft trajectory. The importance of adequate planning for the approach cannot be overemphasized.

The actually available working instruments are summarized and their use is described (Figs. 4 and 5). A very simple self-made instrument is a stiff polyurethane catheter as used in ureteral surgery. They exist with a closed-end tip and a side window. This configuration may be useful for aspiration of cystic lesions. One can easily cut off the tip obliquely to give it the desired sharpness for penetration of tough membranes or capsulas. Such penetration, however, often causes some tearing of small blood vessels and may cause bleeding. Therefore one should never persist too long when penetration is not easy.

Smaller and more flexible catheters, provided with a luer-lock proximal end are useful for aspiration of larger air bubbles, selective rinsing of small areas and

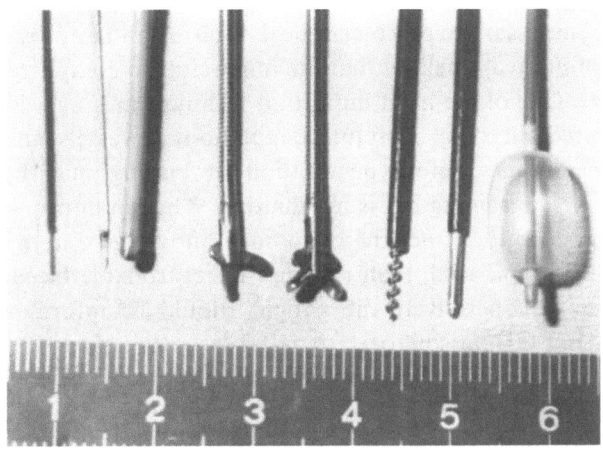

Fig. 4. Available instruments

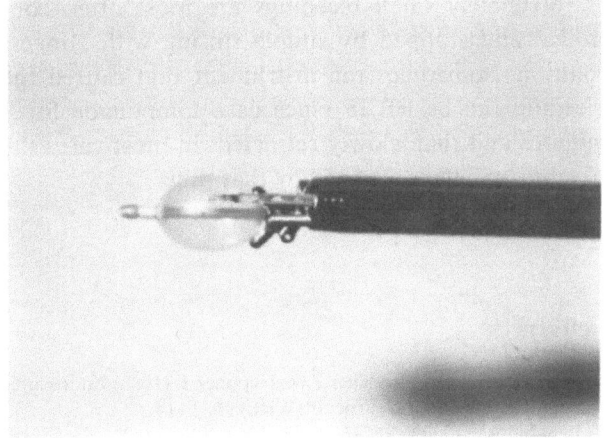

Fig. 5. Fogarty inflatable balloon catheter

aspiration of blood emerging from small vessels so that this blood does not contaminate the whole CSF space.

These small catheters can be introduced through one of the side rinsing channels.

Through the larger working channel several types of grasping forceps can enter and may be used for tumour biopsy, removal of membranes after cutting with the laser, grasping of membranes to stretch them so that other instruments can be applied to them. Although we have no personal experience with the removal of foreign bodies (such as broken pieces of ventricular catheters) this may be possible by means of a grasping forceps. Actually we have four types: one consisting of U-shaped loops so that the grasped object always remains visible; one with sharp teeth to provide a firm grip on the grasped structure, another one containing a cup and particularly suitable for tumour biopsy. Finally a very tiny grasping forceps can be used through one of the rinsing channels as an aid to the use of other instruments through the working channel.

Personally we do not prefer grasping forceps for tumour biopsy since we felt during third ventricular tumour biopsies that the danger of seeding small particles of tumour into CSF spaces is quite real. Therefore we have adapted two excellent and well known stereotactic biopsy instruments to the use through our long endoscope shaft: the Backlund biopsy spiral and the Sedan side-window aspiration needle. Both are inserted deeply in the interior of the tumour so that the danger of seeding tumour cells in our opinion is less than by grasping a specimen at the surface of the lesion with forceps. Moreover capillary bleeding in the tumour is tamponed which is not the case at the surface of the ventricle.

A very useful instrument for this is a small Fogarty inflatable balloon catheter (Fig. 5). It ends bluntly and can already be used to penetrate a thin nonvascular wall. After penetration the balloon can be inflated either inside or below the first small hole, and then retracted to tear a larger opening in a membrane. Its use is limited to about 6–7 mm width; but its advantage is, that it is very atraumatic and does not influence underlying vessels as does the cutting laser. Therefore we prefer the balloon catheter while working very close to vital structures such as the large arteries of the circle of Willis or the basilar artery (for example in perforation of the floor of the third ventricle).

A bipolar coagulation probe is very useful for haemostasis of blood vessels of a surface which has to be penetrated. It requires close contact so that it is suitable for moving aside objects such as the choroid plexus or freely hanging vessels crossing the ventricle. It is very efficient on membranes or septum pellucidum where small vessels have to be coagulated before cutting. The probe consists of an insulated shaft at the tip of which, two separated, naked metal poles provide the current. We use the Malis bipolar coagulator on the 30–35 power index.

Because of its minimal spreading effect, we prefer the bipolar coagulation probe to the laser in those instances where underlying brain structures must not be damaged (for example the thalamus in plexus choroid plexus coagulation).

A bipolar grasping instrument is theoretically attractive but it might be very dangerous when the coagulated vessel is sticking to the forceps as is often seen during open microsurgery.

Tiny microscissors have been tested and nowadays we have two types. The first one has very sharp tips and can be used as a penetrator before cutting. Unfortunately the blades are just straight and sometimes it does not cut but only grasps the tissue. A second type is much more useful. The tips are rounded but one of the blades bears small teeth so that its cutting potency is much higher. Small bridges remaining after laser cutting can very easily be cut off with these scissors, more easily than with the cutting laser itself. While using the scissors the tissue can be stretched by means of a grasping forceps, introduced through one of the rinsing channels.

In the earlier period of our endoscopic work we tried several times to work with the safirtips on a Nd-Yag laser. This experience was very disappointing and we have abandoned this technique. More recently we selected from 6 different new laser types the MBB laser MEDILAS 4060. This offers an excellent coagulating device using about 20–60 as the power range. In the so called "fibertome" mode there is a feed-back control of the temperature at the conical fiber tip (at choice 600°, 700° or 800° C). With the conical sculpted fibers very precise and easy cutting is obtained. All the energy is concentrated at the very tip of the fiber which has to be used in contact for cutting. This offers the advantage that cutting only occurs at the contact site. This offers the advantage that cutting only occurs at the contact side. Coagulation on the

contrary is achieved in the non-contact mode and its effect is modulated both by power (usually ranging from 20 to 40 W) and distance.

Discussion

Endoscopic instrumentation is in full development. Only the systematic application of different tools will enable us to find out what is the most suitable in each specific situations.

Exchange of experience among all different centers using endoscopic techniques will strongly favour progress.

We strongly recommend the use of simple instruments in simple conditions. Endoscopic neurosurgery remains surgery and in spite of the development of computerized robotic technology[2], it will very often be necessary to rely on surgical intuition and improvisation, two qualities that are impossible to computerize. One of the most important technical messages is: avoid bleeding! Careful coagulation of vessels and membranes before penetration or cutting may be time-consuming but is mandatory. When in doubt, it is better to retract the endoscope and to give up an endoscopic trial, than to cause uncontrollable bleeding. Preoperatively the patient should be informed about this possibility.

Personally we are very sceptical about the possibility of coagulating a larger vessel. Bleeding after too extensive removal of an intracerebral haematoma may occur. It is wiser to remove only a critical volume as suggested by Backlund with the use of his haematoma screw.

Fortunately small bleedings are most often controlled and stopped by simple rinsing with Ringers solution. Sometimes the instrument that caused the bleeding can be left in place as a tamponade for 8 minutes and than slowly retracted: in most cases the bleeding will have stopped by that time.

References

1. Auer LM, Holzer P, Ascher PW, Heppner F (1988) Endoscopic neurosurgery. Acta Neurochir (Wien) 90: 1–14
2. Benabid AL, Lavallee S, Hoffmann D, Cinquin P, Demongeot J, Danel F (1992) Potential use of robots in endoscopic neurosurgery. Acta Neurochir (Wien) [Suppl] 54: 93–97

3. Hellwig D, Bauer BL (1992) Minimally invasive neurosurgery by means of ultrathin endoscopes. Acta Neurochir (Wien) [Suppl] 54: 63–68
4. Hor F, Desgeorges M, Rosseau GL (1992) Tumour resection by stereotactic laser endoscopy. Acta Neuochir (Wien) [Suppl] 54: 77–82
5. Karakhan VB (1992) Endofiberscopic intracranial stereotopo- graphy and endofiberscopic neurosurgery. Acta Neurochir (Wien) [Suppl] 54: 11–25
6. Reidenbach HD (1992) Technological fundamentals of endo- scopic haemostasis. Acta Neurochir (Wien) [Suppl] 54: 26–33

Correspondence: J. Caemaert, M.D., Department of Neurosur- gery, University Hospital, B-9000 Ghent, Belgium.

Acta Neurochir (1994) [Suppl] 61: 54–56

Endoscopic Anatomy of the Third Ventricle

T. Riegel[1], D. Hellwig[1], B.L. Bauer[1], and H.D. Mennel[2]

[1] Department of Neurosurgery and [2] Department of Neuropathology, Philipps University Marburg, Marburg, Federal Republic of Germany

Summary

42 cadaver brains in situ were examined endoscopically to work out topographical anatomical landmarks for orientation. The endoscopic route from the chosen precoronal trepanation point to the defined ventricular landmarks has been measured in 22 cases. The identification and measurements of the anatomical landmarks are helpful for safe and atraumatic endoscopical navigation within the ventricular system. Furthermore this article describes and discusses cerebral lesions during ventriculoscopy.

Keywords: Endoscopic anatomy; coronal approach; ventricular system; cerebral midline, neuroendoscopy; ventriculoscopy.

Introduction

Preformed intracranial spaces such as the cerebral ventricles and the subarachnoid cisterns are of particular interest in neuroendoscopy. As we started with our experiments in 1989, we have had problems in spacial orientation and identification of endoscopically visible anatomical structures within the ventricular system. Up to now there has been no endoscopic "cartography" of the ventricular system. For transventricular neurosurgical endoscopic interventions a knowledge of the topographic anatomy of the ventricular system is absolutely necessary.

Methods and Material

The data for the description of the normal endoscopic anatomy within the ventricular system are gained from endoscopic examinations of cadaver brains in situ (Department of Neuropathology, Prof. Dr. med. H.D. Mennel.) So far the total number of endoscopic procedures undertaken is 42. Cadavers with intracranial lesions were excluded from the study.

Technical Equipment

We use flexible and steerable endoscopes with outer diameters of 3.3 to 4.0 mm. Most of them are prototypes which have been developed by Olympus Inc. Tokyo.

Further supplementary instruments are the ultralight camera system OTV-S2-TV camera, the VO-9600 P standard U-matic recorder and the Trinitron colour video monitor PVM 2043 MD (Sony). For documentation and printouts, we use the video printer UV-5000 P (Sony).

For free hand ventriculoscopy, we have designed an endoscopic guiding system (Fig. 1) to reproduce the endoscopic manoeuvres by single perforation of brain substance en route to the ventricular system.

Approaches to the Ventricular System

For ventricular puncture Kocher's trepanation point[4], 2 to 3 cm from the midline, just anterior to the coronal suture is used. This point is ideal for reaching the foramen of Monro and the third ventricle. Other wellknown approaches through Keen's point (temporal) and Dandy's point (occipital) will be used in subsequent studies.

Operation Technique

After burr hole trepanation the dura mater is incised. The fixation screw is inserted into the burr hole and connected with a specially developed "cardanic" joint. The ventricular puncture with the guiding cannula is done in neutral position without horizontal or vertical angle adjustment.

Ventriculoscopy

The ultrathin flexible endoscope (diameter 3.3 mm) is introduced and the guiding tube is pulled back under visual control to the roof of the lateral ventricle and fixed in the "cardanic" device. The endoscope is connected to the camera system and the ventriculoscopy is transmitted to the video-unit.

Results

The coronal approach permits endoscopy of the frontal horn including the central part of the lateral ventricle, the foramen of Monro and the third ventricle. Furthermore it is possible to reach the aquaeduct of Sylvius, going further to the fourth ventricle.

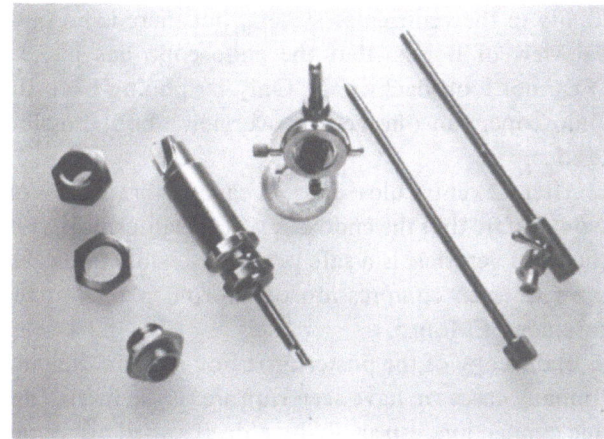

Fig. 1. Endoscope guiding system with the cardanic joint

Topographic Landmarks

Lateral Ventricle

From the *coronal approach* one reaches first the central part of the lateral ventricle near the frontal horn. The frontal horn can be distinguished by the lack of choroid plexus. The lateral wall is formed by the caput ncl. caudatus with subependymal veins; medially one finds the septum pellucidum with septal veins. The choroid plexus and the foramen of Monro serve as landmarks for the central part of the lateral ventricle. The plexus is situated in the floor of the lateral ventricle, the thalamostriate vein is located laterally

and medially the confluence of septal veins. These three structures form a typical Y-shaped configuration[1] which is ideal for orientation (Fig. 2). The foramen of Monro is anteriorly and laterally formed by the fornix and the posterior border is built by the anterior tubercle of the thalamus[5]. After angling the flexible endoscope, one locates the central part of the lateral ventricle up to the ventricular trigone. Laterally one finds the corpus nuclei caudati with the thalamostriate vein under the lamina affixa and the medial border is formed by the fornix and the septum pellucidum.

Third Ventricle

Going through the interventricular foramen, one enters the anterior part of the third ventricle. The mamillary bodies glimmer through the thin ventricular base. This membrane, which is the posterior part of the tuber cinereum, is the location for performing third ventriculostomy[2] (Fig. 3). The infundibular recess, the optic chiasma and recess are visible in the frontal part. Laterally we find important hypothalamic regions and also parts of the fornix.

In the posterior part of the third ventricle, the first visible structure may be the interthalamic connection, but this is absent in 25% of cases[3]. Going further posteriorly under the massa intermedia, one reaches the pineal region. The floor is made up of the poste-

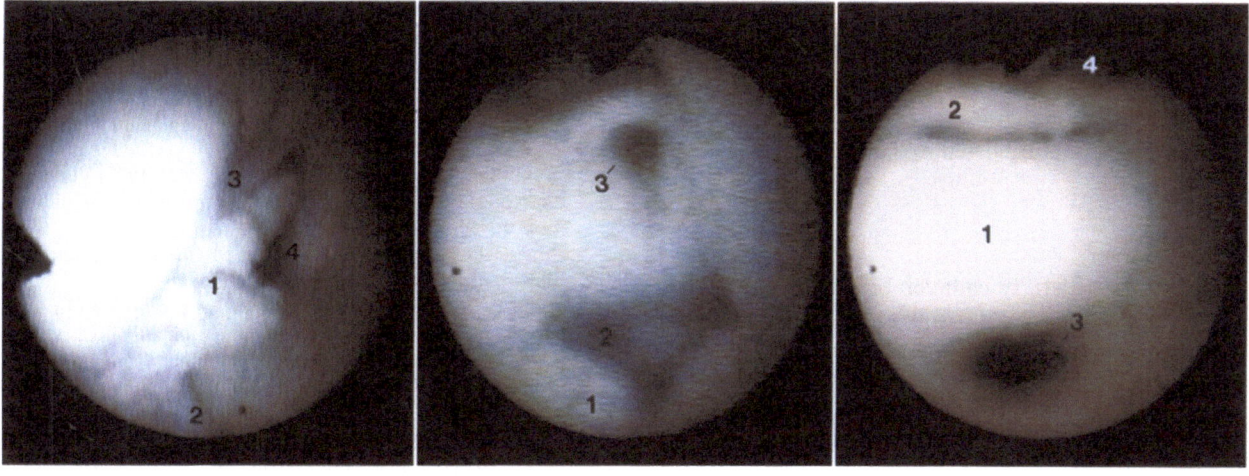

Fig. 2.   Fig. 3   Fig. 4

Fig. 2. Typical Y shaped configuration in the region of the interventricular foramen. *1* choroid plexus, *2* thalamostriate vein, *3* septal vein, *4* foramen of Monro

Fig. 3. Localization for third ventriculostomy. *1* mamillary bodies with *2* premamillary membrane and *3* infundibular recess

Fig. 4. Posterior wall of the third ventricle. *1* posterior commissure, *2* habenular commissure, *3* aqueduct, *4* suprapineal recess

Table 1. *Topographic Measurements*

Landmark	Median	Range
Roof of the lateral ventricle	4.50 cm	3.5–5.3 cm
Foramen of Monro	6.10 cm	4.6–6.9 cm
Mamillary bodies	7.70 cm	7.0–8.4 cm
Optic recess	8.80 cm	7.7–9.7 cm
Infundibular recess	8.50 cm	7.5–9.6 cm
Posterior commissure	8.90 cm	7.8–9.7 cm
Pineal recess	9.50 cm	8.4–10.8 cm
Aqueduct of Sylvius	9.30 cm	8.2–10.0 cm

rior perforated substance and the uppermost mesencephalic tegmentum. On each side, the third ventricle is bounded by the medial surface of the thalami[5].

The posterior commissure separates the pineal recess and the habenular commissure from the entrance to the aqueduct (Fig. 4) in the small posterior wall. Reaching the suprapineal recess two longitudinal strips of choroid plexus and branches of the internal cerebral vein project toward the roof of the third ventricle. The roof is formed by the tela choroidea. Below the posterior commissure one finds the entrance of the aqueduct. From there, it is not difficult to reach to the fourth ventricle.

Endoscopic examinations of the fourth ventricle are in preparation.

Topographic Measurements

The endoscopic way from the choosen precoronal trepanation point (outer table of the skull) to the defined ventricular landmarks were measured in 22 cadavers with a mean age of 69 years. Sex distribution was 10 females and 12 males.

The main data are listed in Table 1.

Discussion

We decided to study cadaver brains in situ with the idea of simulating the operative procedure in physiological relationships. The brain is without any fixation and the ventricular system is still filled with CSF. The vessels are sometimes collapsed and autolytic processes may impair optic conditions, but this model gives an excellent overview for endoscopic navigation. The ultrathin flexible endoscope allows one to observe and estimate the neuro-anatomical structures with little trauma. The flexibility provides good manoeuvre-

ability in the ventricular system, but there is no optical view of lesions that the endoscope has passed. It cannot look backwards. Only by pulling back the endoscope, can one get an overview about possible lesions.

After 42 ventriculoscopies in cadaver brains in situ, we can state that the endoscopy of the anterior part of the third ventricle is a safe procedure. Only in 2 cases have we seen compression of choroid plexus in the foramen of Monro.

Endoscopy of the posterior region is more difficult. In many cases we have seen rupture of the interthalamic connection, especially by a deep seated adhesion. In 7 cases we could find lesions of the anterior wall of the foramen of Monro, which is formed by the fornix.

In all efforts to reach the fourth ventricle, we have seen lesions of the interthalamic connection and the area of the foramen of Monro.

As a conclusion we can agree, that endoscopy of the third ventricle is a safe and atraumatic procedure within the meaning of a minimal invasive technique, but the effort to reach the fourth ventricle from Kocher's trepanation point may be coupled with lesions of functionally important structures.

The identification and measurements of the anatomical landmarks is helpful for the assessment of dimensions and spacial orientation. With the aim to minimize brain tissue trauma and to reproduce endoscopic procedures and measurements in every single case, we have designed the endoscopic guiding system with the cardan joint.

References

1. Fukushima T, Ishijima B, Hirakawa K, *et al* (1973) Ventriculofiberscope: a new technique for endoscopic diagnosis and operation. J Neurosurgery 38: 251–256
2. Jones RFC, Stening WA, Boydam M (1990) Neuroendoscopic third ventriculostomy. Neurosurgery 26: 86–92
3. Lang J (1992) Topographic anatomy of preformed intracranial spaces. In: Bauer BL, Hellwig D (eds) Minimally invasive neurosurgery—MIN. Acta Neurochi (Wien) [Suppl] 54: 1–10
4. Mapstone TB, Ratcheson RA (1985) Techniques of ventricular puncture. In: Wilkins RH Rengachary SS (eds) Neurosurgery. McGraw-Hill, New York, pp 151–152
5. Netter FH (1983) Nervous system, Vol 1, Part 1. Anatomy and physiology. Ciba Collection of Medical Illustrations, pp 30–31

Correspondence: Thomas Riegel, M.D., Department of Neurosurgery, Philipps University Marburg, Baldingerstrasse, D-35033 Marburg, Federal Republic of Germany.

Acta Neurochir (1994) [Suppl] 61: 57–61

Endoscopic Anatomy of the Ventricles

K.D.M. Resch, A. Perneczky, M. Tschabitscher[1], and St. Kindel

Department of Neurosurgery, University of Mainz, Federal Republic of Germany and
[1] Department of Anatomy, University of Vienna, Austria

Summary

The endoscopic view offers a new anatomical dimension to the neurosurgeon. The fact makes it basically necessary to study the topographic anatomy under endoscopic conditions. In this paper attention was drawn to the ventricles because they are the most common region of clinical application. In 25 specimens neuro-endoscopic explorations of the ventricles have been done. The dissections have been carried out through one- and two burr hole approaches (two working endoscopes at the same time). The instrumentation includes rigid 4 mm and 6 mm endoscopes. The procedures have been documented by continuous video recording and parallel photography.

Keywords: Neuroendosopy; ventricular system; anatomy.

Introduction

Minimal invasive techniques are not new in neuro-surgery but endoscopes are the tools necessary to achieve key hole surgery in selected cases. The general road maps of endoneurosurgery did not exist. A neuro-endo-anatomy is therefore necessary and the rule should be that the anatomy is ahead of surgery. As Yasargil mentioned, while neurosurgery changed from macro- to microtechnique, microneurosurgery is not just a question of the technical tools[15]. There are some fundamental differences to microsurgical anatomy; we see all as in a "fish eye aspect" with a broad angle view and perspective, when we reach parts that were optically hidden to the microscope. We can observe ipsilateral preparation from the contralateral side. We can bring the light optimally to the deepest points and come as near to the structures as we need. The topographical overview might be zero, neighbouring structures move out of view while going deeper, orientation is difficult, rotation of the picture is easily possible and might not represent the true position; this may lead to mistakes. At present strategies of safety and variety of instruments are not de-veloped enough and use is therefore very limited. These facts show the importance of a safe anatomical basis for neuro-endoscopy.

Material and Methods

25 cranial endoscopical explorations have been completed and results documented by photos and video.[10] The first three explorations and endoscopy seminars have been done with plastic specimens, which are preserved by a plastic impregnation technique[4–8,11,13]. The others were done in non-fixed specimens.

The technical equipment comprises: Rigid endoscopes 4 mm 0′, 30′, 70′ and 6 mm 5′ with canals for irrigation, suction and instruments, photo-video system and surgical instruments.

Regions of exploration were the ventricles. Approaches were microsurgical (open endoscopy), double burrhole and burrhole in air and water. In some cases a video-endo-dissection technique was used where the endoscope is used as a substitute for the microscope and the view is displayed on the monitor during endoscopy.

In this presentation attention is focussed on the ventricles because they are commonly the first area of clinical application.

Results

To use the endoscope intracranially is to enter a new world with several difficult conditions: The first difference to the macro- and micro world are the optical conditions which is a "fish eye view" (Figs. 2 and 3). Moreover there is a very different perspective, that leads to the effect that the size of a structure changes markedly with the distance of the lens. A tiny vessel near to the lens might look double the size of the main vessel. The second difference is a new eye-hand-co-ordination in a medium with the highest necessity of safety in guiding, orientation and manipulation. This leads to a stereotactic situation in planning and guid-ing. Of course it is possible to manoevre the scope freehanded, but this may lead to damage.

The main advantage of the rigid scope is its optical

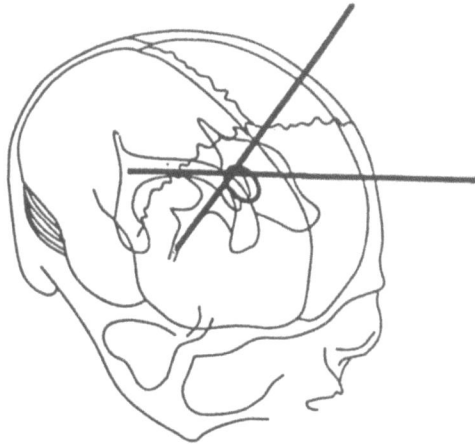

Fig. 1. The direction of a rigid scope has to be planned in order to reach different parts of the ventricular system

brilliance compared to fiberscopes and the exact guiding conditions. The main disadvantage is the difficulty of looking around corners. If using the rigid scope with small deviation movements and with an angled view to compensate for the missing flexibility it is of great importance to control the movements of the whole scope. All structures that have been passed are no longer visible and therefore in danger. This leads to the need for a backward view or a back situated control view for the moved scope. The second, controlled and movable scope could also be a fiberscope used through one of the canals of the rigid scope.

More than in micro-approaches the endo-approach (burrhole) must be planed exactly. The variations of position are numbers and every location has different possibilities and difficulties.

The classical burrhole position for the lateral ventricle is the precoronal midpupillary point. To get into dorsal parts of the lateral ventricle the direction is more dorsal as if to pass through the foramen of Monro (Fig. 1). To enter the rostral third ventricle the burrhole position is more dorsal. To enter the dorsal third ventricle and in selected cases the aqueduct

and fourth ventricle the position of burrhole is more frontal.

One of the main structures of orientation in the frontal horn of the lateral ventricle is the choroid plexus that leads laterally accompanied by the lamina affixa, the thalamus, the thalamostriate vein and the nucleus caudatus to the dorsal parts of the ventricle (Fig. 4). The thalamostriate vein often is covered by the tissue of stria terminalis.

The other main structure is the foramen of Monro that leads to the third and fourth ventricle (Figs. 5 and 6). The foramen of Monro is surrounded by the column of the fornix and thalamus, the plexus sharply curves medial and runs along the roof of the third ventricle. In hydrocephalic conditions the foramen of Monro is widely open and gives a view of the floor of the third ventricle (Figs. 5 and 6).

The main structure is the typical figure of mamillary bodies and the rostral pointing infundibular recess. Lateral the hypothalamus forms the wall and the floor leads to the aqueduct. In many cases there is no interthalamic connexus (Figs. 5–7).

The main structure of the roof anteriorly are the fornices, the anterior commissure and the choroid plexus (Figs. 8 and 9), posteriorly there is the entrance to the aqueduct, the habenular commissure with the pineal gland and the posterior commissure (Fig. 10).

In hydocephalic conditions it is possible to reach the fourth ventricle even with a rigid 4 mm-scope. The main dorsal structure is the plexus of the fourth ventricle fixed at the transparent tela choroidea through which the cortex of the cerebellum is visible. The plexus leads dorsocranially to the fastigium and laterally on both sides to the foramen of Luschka in the lateral recess (Fig. 11). The main ventral structure looks like the quadrigeminal plate through the endoscope but is the medial eminentia superiorly and facial colliculus inferiorly on both sides divided by the median sulcus of fourth ventricle. Beneath the medullary striae can

---

Fig. 2. Interpeduncular cistern seen through the operating microscope

Fig. 3. Interpeduncular cistern seen through a rigid endoscope 4 mm, 30. Note the "fish eye" effect

Fig. 4. Right lateral ventricle with foramen of Monro, choroid plexus and lamina affixa

Fig. 5. Foramen of Monro formed by the fornix and the thalamus. The plexus curves sharply towards the roof of third ventricle. In the depth the mamillary bodies are seen

Fig. 6. Through the foramen of Monro the floor of third ventricle is visible. Mamillary bodies are easily recognized, the dark triangle between them is often a membrane that can be penetrated into the interpeduncular cistern (see Fig. 3)

Fig. 7. The floor of the third ventricle showing mamillary bodies with a small membrane between them and the infundibulum

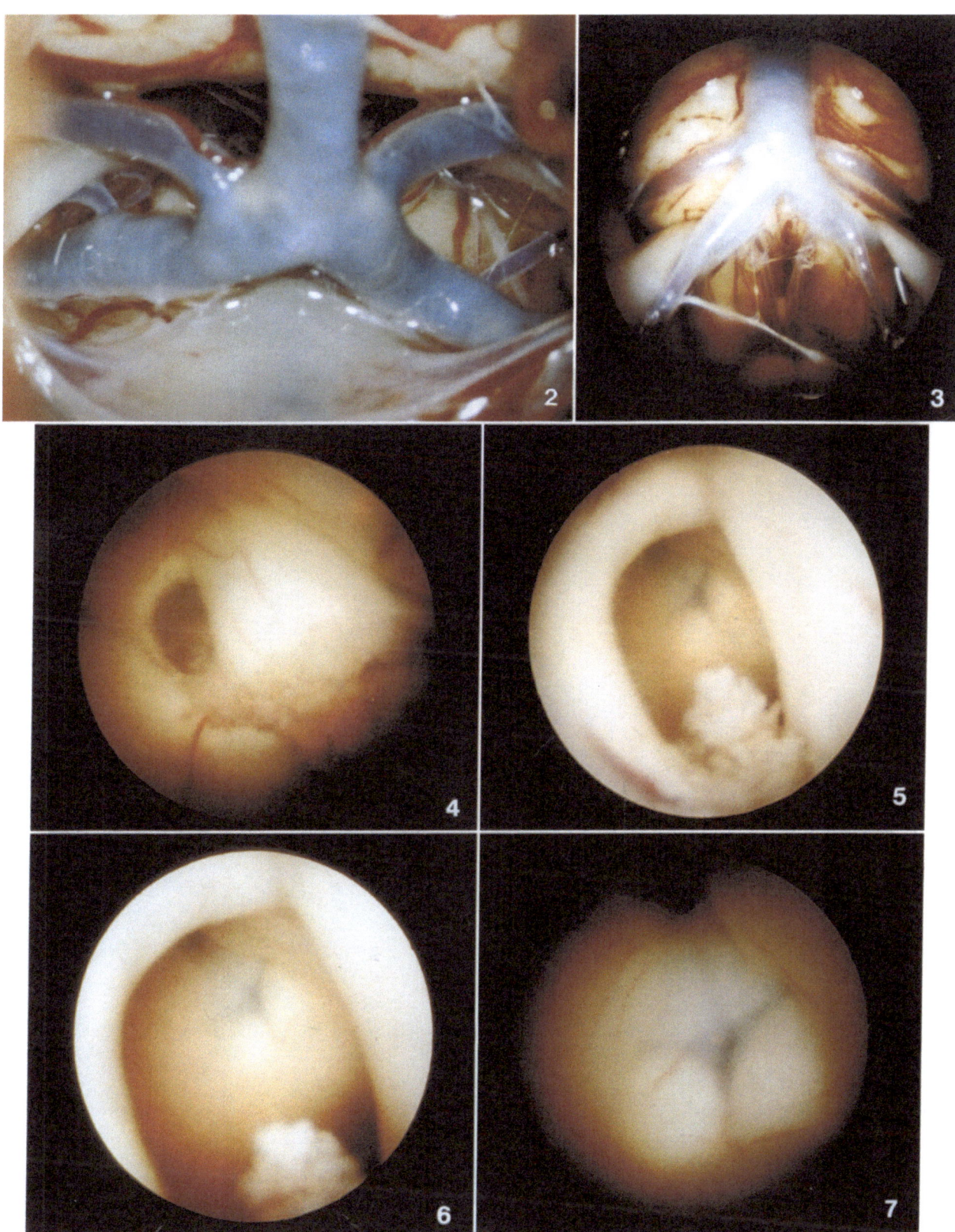

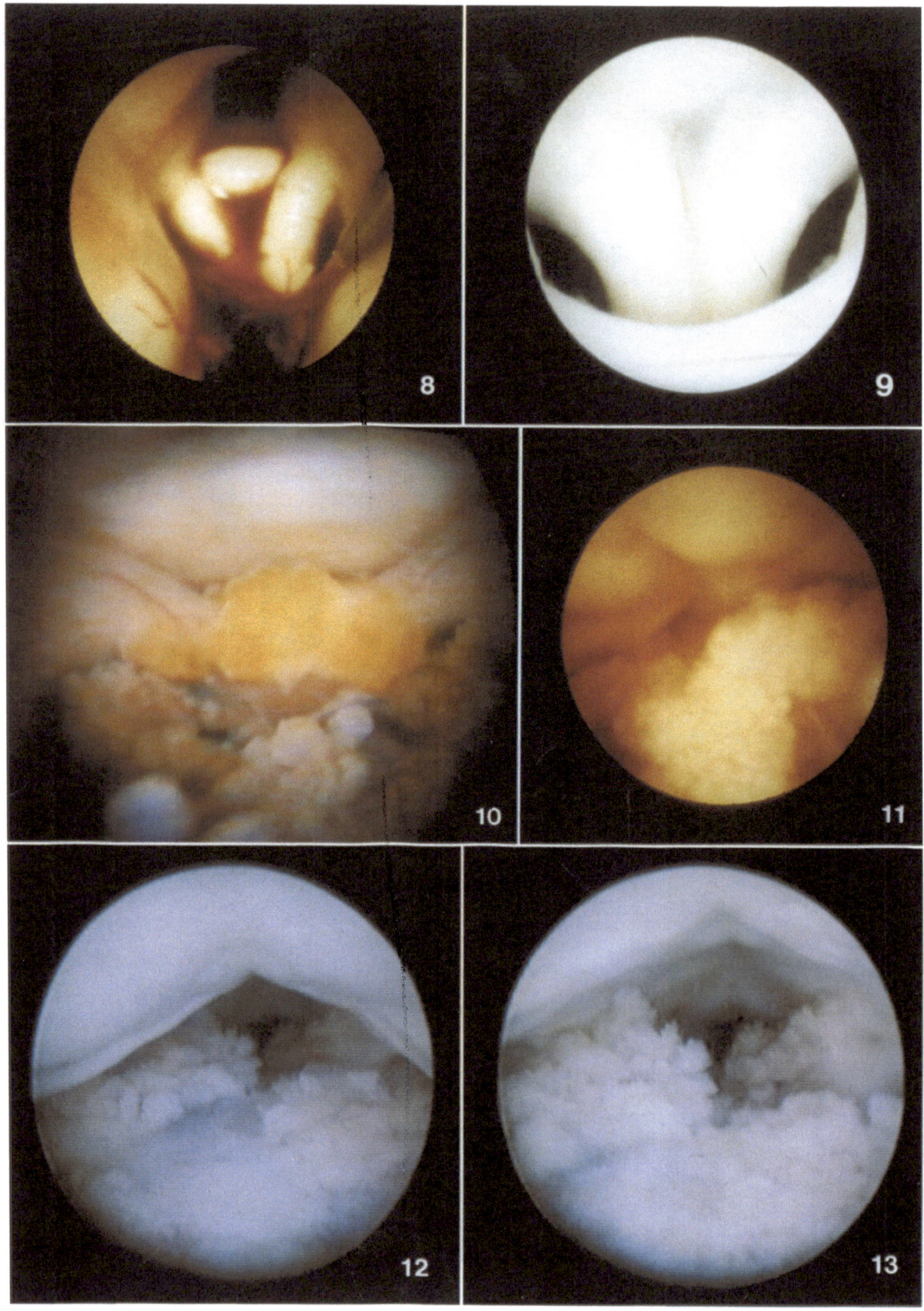

usually be identified and laterally the vestibular region (Figs. 11 and 12).

Inferior to the plexus and the medullary striae (dorsal acoustic striae) the caudal fourth ventricle begins and in the depth the foramen of Magendi can be seen. There the ventral wall of the ventricle is formed by the hypoglossal and vagal triangles on both sides (Figs. 12 and 13).

Discussion

Though neuro-endoscopy was used early by Espinasse in 1911[3], neurosurgery lags behind other disciplines in surgical endoscopy[1]. This is due to the absolute necessity of secure intracranial working. On the other hand endoscopy seems to be an adequate tool to work in a space like the ventricle. The main reason is the atraumatic approach, really entering through a key hole. Neuroendoscopy seems to be the logical development from a key-hole-microsurgical point of view.

Of course neuroendoscopy will not easily become popular[9] if the roadmaps of neuro-endoscopy are not available. As can be analysed respectively microneurosurgery had the same problems originally and Yasargil consistently draws our attention to the point that the problems are not only technical, but start with anatomical difficulties and concepts[15].

Today we see the chance to practice a classical approach: that anatomy should precede surgery. But it has not to be an anatomy of statistics. Surgeons need to be guided by clinical anatomy as Tandler pointed out[14]. Therefore we need the description of a "gestalt-anatomy"; the clinical anatomist has to use the surgical technique at the dissection table[12]. The best method to train and study neuro-endo-anatomy is the non-fixed specimen in the first two days post mortem. One should start in very old specimens with hydrocephalic conditions. Young specimens or even brain oedema might result in frustration. The work is time consuming but the effort of training and the results are rewarding.[10]

Topographical and microsurgical anatomy is indispensible. At the present time there are no puplications for real discussion but the message is that: "To see is to understand"[2].

References

1. Buess G (ed) (1990) Endoskopie. Von der Diagnostik bis zur neuen Chirurgie. Deutscher Ärzteverlag Köln
2. Clark K (1969) Leonardo da Vinci. Rowohlt, Reinbek
3. Griffith HB (1986) Endoneurosurgery: endoscopic intracranial surgery. In: Symon L et al (eds) Advances and technical standards in neurosurgery, Vol 14. Springer, Wien New York, pp 3–23
4. Hagens Gv (1985/86) Heidelberg plastination folder. Anatomisches Institut I, Universität Heidelberg, Federal Republic of Germany
5. Hagens Gv, Tiedemann K, Kriz W (1987) The current potential of plastination. Anat Embryol 175: 411–421
6. Hagens Gv (1979) Impregnation of soft biological specimens with thermosetting resins and elastomers. Anat Rec 194: 247–255
7. Hagens Gv, Whalley A, Maschke R, Kriz W (1990) Schnittanatomie des menschlichen Gehirn. Steinkopff, Darmstadt
8. Hagens Gv, Romrell LJ, Ross MH, Tiedemann K (1990a) The visible human body. Lea and Febiger, Philadelphia
9. Heilmann CB, Cohen AR (1991) Endoscopic ventricular fenestration using a "salin torch". J Neurosurg 74: 224–229
10. Perneczky A, Tschabitscher M, Resch KDM (1993) Endoscopic anatomy for neurosurgery (Atlas). Thieme, Stuttgart
11. Resch KDM (1989) Use of plastinated specimens in the demonstration of microsurgical approaches to the cranial base. Third International Conference on Plastination. San Antonio, Texas 1986. J Int Soc Plastination 3:28–33
12. Resch KDM (1990) Beitrag zur Zugangsanalyse und zum Zugangsdesign des transoral-transpharyngealen Weges zum Hirnstamm. Dissertation, Universität Heidelberg
13. Resch KDM, Perneczky A (1992) Use of plastinated craniums in neuroendoscopy. J Int Soc Plastination 6: 15–16
14. Tandler J (1929) Lehrbuch der systematischen Anatomie, Bd 4. Vorwort. Vogel, Leipzig
15. Yasargil MG (1984) Microneurosurgery, Vol 1. Thieme, Stuttgart

Correspondence: K.D.M. Resch, M.D., Neurochirurgische Klinik und Poliklinik, Universität Mainz, Langenbeckstr. 1, Gebäude 505, D-55101 Mainz, Federal Republic of Germany.

◁————————————————————

Fig. 8. The roof of third ventricle in a case of missing interthalamic connexus (in air). The fornices and the anterior commissure, above the lamina terminalis, lateral the thalamus forming the foramen of Monro with the fornix on both sides, beneath the bleeding choroidal plexus

Fig. 9. The roof of third ventricle in a case with an interthalamic connexus in water. Both fornices and open foramina of Monro, above the anterior commissure beneath the interthalamic connexus

Fig. 10. The upper fourth ventricle shows ventrally the typical structure of the four colliculi which are eminentia medialis and beneath the colliculus facialis on both sides. Dorsally is the T-shaped plexus

Fig. 11. The lower fourth ventricle is beneath the plexus and the dorsal acoustic striae showing ventrally the triangle of hypoglossal and vagal nuclei dorsally the foramen of Magendi

Figs. 12 and 13. The lower fourth ventricle is beneath the plexus and the dorsal acoustic striae showing ventrally the triangle of hypoglossal and vagal, nuclei dorsally the foramen of Magendi

Acta Neurochir (1994) [Suppl] 61: 62–68

"Stereology" of Intracranial Lesions

H.D. Mennel and **C. Roßberg**

Department of Neuropathology, Medical Center of Pathology, Philipps University Marburg,
Federal Republic of Germany

Summary

Endoscopy of the intracranial space requires a new understanding of the anatomy and pathology of pertinent structures. This meets with the new development of imaging methods which equally require three dimensional interpretation of intracranial pathology. The stereological arrangement of intracranial lesions is examplified on three neuropathological conditions: brain tumours, territorial infarction and mass displacement.

Keywords: Three dimensional intracranial pathology; brain tumours; brain infarction; intracranial mass displacement.

Introduction

Questions of stereology, i.e. the spatial arrangement of lesions within the intracranial cavity are essential topics of neuroanatomy and neuropathology: Localisation of function can be deduced from the observation of the site of brain damage. One of the outstanding examples is the identification of the motor speech area by Paul Broca in a traumatic brain[2].

Localisation of higher mental and psychological functions has a long and remarkable tradition. Greek, islamic and medieval medicine localized psychological and sensory capacities—memory, taste, olfaction and vision—within the different parts of the ventricular system[1,3]. Progress of clinical neurology and neurosurgery in the 18th century was only possible after the detection of basic phenomena of the structure of the nervous system, such as pyramidal tract crossing by Gall *et al.*[12]; the next generation of clinicians and neuropathologists such as Theodor Meynert, Emil Paul Flechsig and Bernhard von Gudden initiated the era of neurohistology and thereby opened new insight into neurological function and its localisation[7].

Emil Kraepelin by founding the "Deutsche Forschungsanstalt für Psychiatrie" and assembling there such distinguished neuropathologists as Franz Nissl, Aloys Alzheimer and Walter Spielmeyer challenged neuropathology with the task to identify the morphological basis of mental function and dysfunction[6]. Although the major psychoses could not be unravelled by neuropathological methods, the efforts were not in vain: Neurological diseases such as systemic atrophies and demyelinations were pathogenetically defined and explained[15]; neuropathological features of irreversible organic psychosis could be described which turned out to be very promising for further research e.g. in Alzheimer's disease[4].

Two techniques were created by neuropathology in answer to questions of localisation of function, which were true forerunners of modern methods: Special stains in neurohistology depicted cellular constituents of the nervous system such as neuronal perikarya, axons and myelin sheeths as well as glial cells and fibers. The same structures today are visualized more clearly and more reliably by immunocytochemistry. The second answer consisted in performing double hemispheric preparations. These neurohistologically stained whole brain sections provided an excellent tool to study connections and lesions between different parts in a topographical manner. This technique anticipated modern imaging methods, in which structures and lesions are shown in a similar way. Much of the accumulated findings of neuropathology have been reappraised during recent years; disease entities known only to the small group of people examining professionally postmortem brains became—probably prematurely—very frequent diagnoses, for instance Binswanger's disease. Yet, such a reappraisal revealed the tremendous wealth of knowledge gathered by classical neuropathology. We shall consider here three of the most frequent pathological intracranial conditions: Tumours, infarction and mass displacement.

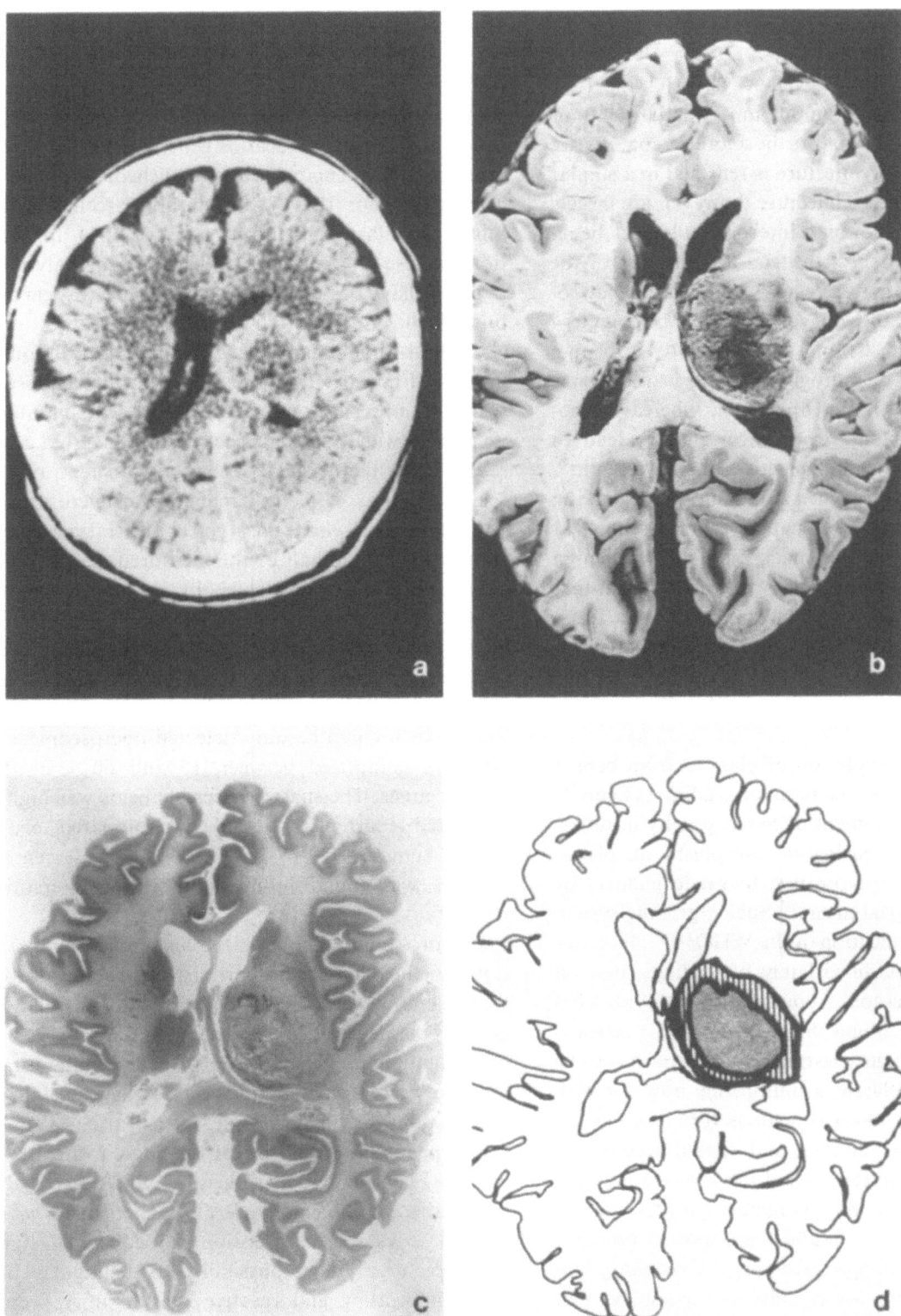

Fig. 1. Comparison of CT scan (a), cut brain slice (b), histological double hemispheric preparation (c) and schematic drawing (d) for morphometric analysis in a thalamic glioma. In this case the necrosis is central. Histology shows that its extention is larger than suggested by CT-scan and naked eye observation

Stereology of Intracranial Tumours

Imaging methods, such as CT-scans, MRI and PET can give a picture of a tumour in its spatial relations to the surrounding brain. Furthermore, in some cases, the internal tissue architecture is reflected in a similar way. Cell and tissue architecture, however, are largely correlated to the tumour's internal biological behaviour. This holds especially true for tumours of the glioma group. Aggressiveness and malignancy of these tumours are directly correlated to the degree of pleomorphism of cells and tissue. Relatively benign "isomorphic" oligodendrogliomas and astrocytomas tend to recur years after operation, whereas "pleomorphic" glioblastomas multiforme mostly lead to the patient's death within the period of a few months[8]. This increasing heterogeneity of tumours depending on their increasing malignancy is represented at different levels: On the macroscopic one, benign tumours tend to be rather homogeneous, whereas malignant tumours have distinct compartments. These compartments can be seen in postmortem preparations of tumour bearing brains, in stained double hemispheric sections as well as in imaging methods (Fig. 1).

The continuous evolution of gliomas from benign to malignant states is the basis of grading systems[10]. The intrinsic development of heterogeneity in different parameters (increasing pleomorphism, anaplasia, variegation) makes it possible to grade gliomas by purely morphological means[9]. Such a grading system underlies the classification of the WHO[16,17]. Its rationale has been underlined lately by findings, that on the molecular level too, a continuous evolution from relatively benign glioma through anaplastic astrocytoma or glioma to glioblastoma multiforme has to be postulated[5]. Glioblastoma multiforme however with its well delineated compartments is very much suited for morphometric handling and spatial reconstruction. We have analyzed several intracranial tumours —amongst them twelve malignant glioblastomas— morphometrically on double hemispheric preparations. In all of the glioblastomas, CT scanning had been performed previously, once or repeatedly. Yet, a direct comparison of measurements of stained sections and CT-scans was not attempted, since we deal with different resolution levels. Five of the patients with glioblastoma had been treated by various methods (operation, interstitial and exogeneous radiation, chemotherapy) and seven were entirely untreated.

Clinical data in all patients were such, that advanced tumour growth or recurrence was obvious. Measured tumour areas ranged from six to 45 cm^2, which corresponds roughly to a volume of 25 to 150 cm^3. All tumours had signs of local or general pressure. Three gross compartments were distinguished: Brain adjacent to tumour (BAT), proliferating (cellular and highly vascular) zone and necroses (Fig. 1). In addition to these gross features, a distribution analysis of size and shape of tumour cells and vessels was carried out, which is not reported here[11].

There is no constant growth pattern, although growth with nearly round or slightly irregular border seemed to prevail. Form factors of the whole tumour areas were widely scattered between near zero, i.e. no recognizable figure and 1, i.e. round. In fact, multifocal as well as perfectly circular forms were found. Round tumours had the typical central necroses; slightly more than half of the measured necrotic foci belonged to this type, which by its large extent is mostly well depicted in the CT-picture. The remaining were either excentric and scattered necroses (Fig. 2) or the elongated necroses highly characteristic for glioblastoma, with and without palisading of cells around them, which could be only detected microscopically. Necroses comprised between 15 and 60 % of the tumour areas. The share of necrotic fields was higher in treated recurrent (40 +/– 11 %) compared to untreated tumours (29 +/– 17 %). The largest central necrosis was found in a tumour after interstitial irradiation.

The proliferation zone in the histological preparation is defined as the compartment with recognizable living tumour cells. In the CT scan, its correlation is the enhancing highly vascular proliferating rim or garland figure. Separation of this proliferating/vascular area from necrotic foci is easy in stained preparations; in "imaging" pictures, clear cur distinction depends upon the size of the compartments. In contrast, central and peripheral boundaries of the brain adjacent to tumour (BAT) compartment were difficult to establish: extension of oedema in computed pictures and impossibility to distinguish between infiltrating and reactive cells in microscopic sections made it feasible to define this compartment broadly.

Newformed vasculature was most prominent in the described proliferating/enhancing rim. Large sinusoidal and glomerular formations were most frequent. Less of both varieties were observed in recurrences

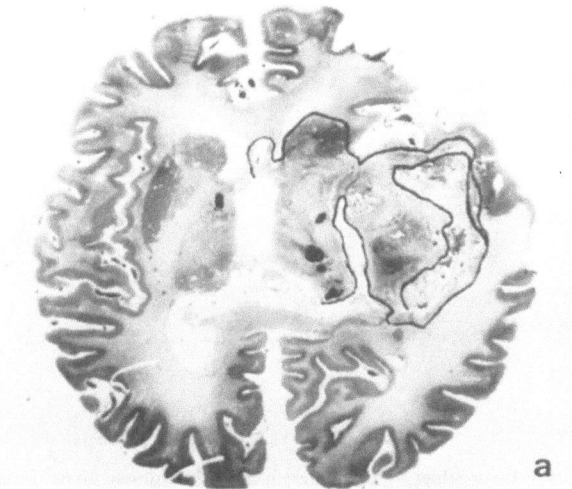

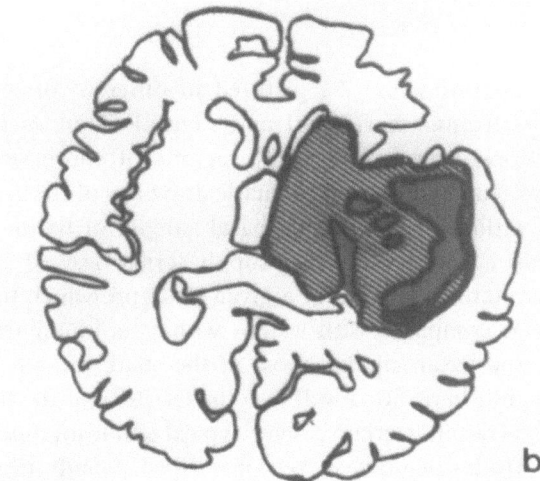

Fig. 2. Large hemispheric glioblastoma with small foci of central necrosis (islands in schematic drawing) and larger peripheral necrosis (hatched areas). In this case, the microscopical visible necrosis is peripherally situated. (a) Double hemispheric stained preparation. Compartments are marked and shown again in (b) schematic representation

Table 1. *Mean Vascular Density in Different Compartments*

Brain adjacent to tumor (BAT)	3.5%	range 1.8– 6.1
Proliferating rim	8.9%	range 3.9–20.7
Necrosis	13.7%	range 5.0–26.2

course overlapping features in intermediate forms have to be kept in mind, when a tumour is approached by stereotactic or endoscopic means. One has in addition to be aware, that imaging pictures are very helpful as a guide for such a procedure, but are by no means completely consistent with the true morphological condition represented by the microscopically analyzed tissue and cell composition.

Spatial Arrangement of Arterial Infarction

Vascular incidents are even more frequent than intracranial tumours and pose similar difficult problems as to their mangement. Infarction of large supratentorial and brain stem arteries cause known neurological syndromes, often resulting in death or disability. Preventive and therapeutic strategies are in continuous discussion. Typical infarctions are recognized, which depend on vascular territories, on extra- and intracranial anastomoses as well as on the pathogenesis of the hypoxic damage[18]. In this field too, the introduction of CT methods into clinical neurology has revived older concepts and findings. Imaging methods again clearly visualize the different extensions and shapes of hypoxic lesions, which by comparison with histological double hemispheric preparations are immediately evident. For the better understanding of those lesions however, the representation of vascular territories is extremely helpful as some kind of intermediary missing link between pathogenesis and morphology. Postmortem angiography can be performed to this end before cutting and processing brains in the usual neurohistological procedures (Figs. 3–5).

Postmortem angiography, when performed on the large supra- and infratentorial arteries and radiologically analyzed on whole brains, gives a puzzling picture of the stereological arrangement of these vessels. In brain slices analogous to CT planes, anatomical and pathological vascular distribution in postmortem angiographic pictures becomes quite clear. But in the sagittal plane too, a whole territory may be marked with its clear boundaries as in the case of the superior cerebellar artery (Fig. 3). Some pathological lesions are even better delineated by angiography than in spe-

after therapy. Vascular density, i.e. the number of counted vessels per area steadily increased from the BAT-compartment through the proliferating zone towards the necrotic center, where it reached its higest score, when measured as outer surface. Equally, the relative share of vessels per measured area increased from the periphery to the center (Table 1). Vessels have been counted regardless of whether or not they were functionally active or inactive, i.e. closed by thrombotic material as is generally found in necroses.

These arrangements of tumours in either monomorphous fields in the case of most benign and delineated compartments in malignant specimens with of

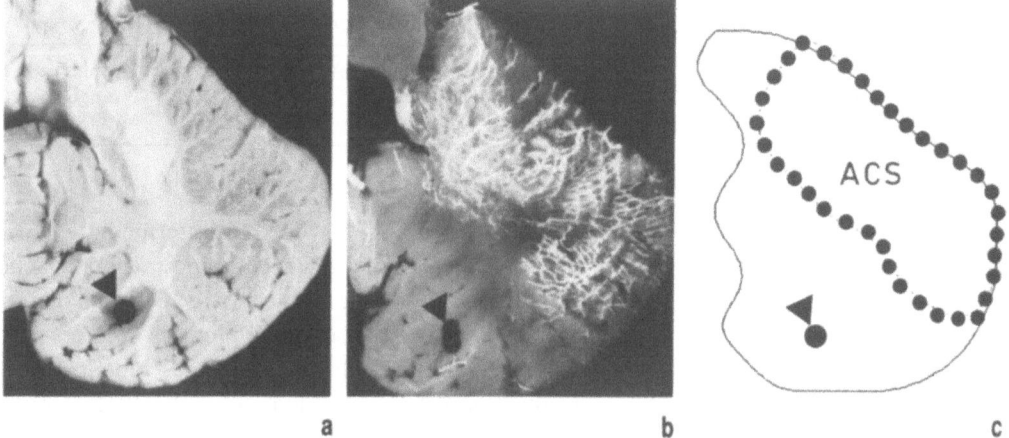

Fig. 3. (a) Horizontal plane showing cerebellar hemisphere. (b) Representation of superior cerebellar artery of identical plane. Small haemorrhage serves as orientation mark (arrowhead). (c) Scheme of the supply area of the superior cerebellar artery

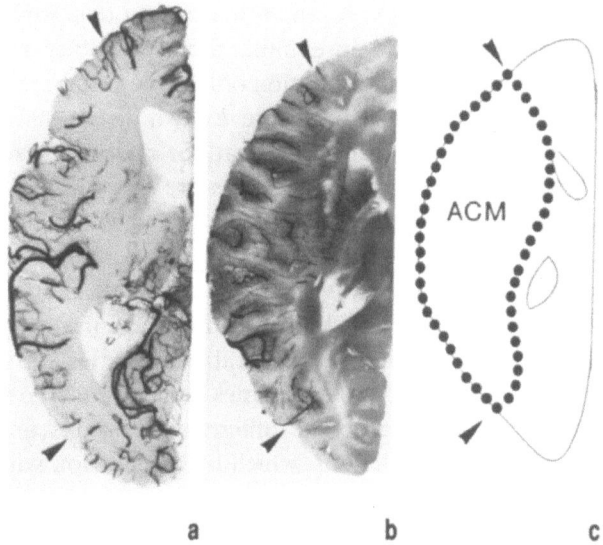

Fig. 4. (a) Postmortem angiography with reduction of small intracerebral vessels in the area of the middle cerebral artery. (b) "Normal" selective postmortem angiographic picture of the territory of this vessel. (c) Drawing of the supply area of this brunch of the middle cerebral artery

cially stained sections (Fig. 4). Necrobiotic changes leading to altered consistency, mass displacement, compression of small vessels and microscopic changes in cytology and tissues can be compared on almost identical planes. This comparison is especially valuable in small infarcts of brain stem and basal ganglia. But since they present with only minor, difficult and slight symptoms, they were neglected and rediscovered by imaging methods.

For postmortem angiography of such small vessels, selective contrast medium injection is preferable. Sup-

ply territories may be pictured in different orientations, frontal, horizontal and sagittal: Thus, a stereoscopic picture of supply regions of those small vessels might emerge. Schematic drawings of the areas e.g. within the confines of basal ganglia of the brain yields a "map of vascular supply territories" for the three standard planes of a given structure which then can be compared with lesions within its boundaries. For the basal ganglia, most of the small infarcts by size and shape fitted well into the territory of the (lateral) striatal arteries[13]. Very typical small infarcts in the frontal plane have been described equally in the territory of the posterior thalamoperforating artery (Fig. 5). These infarcts may by the variability of the vascular tree even occur bilaterally in a very similar manner[14].

Spatial computerized reconstruction however of the vessel course, the lesion and the anatomical structures allows one to examine the interrelations of the different parameters. For the territory of the (lateral) striatal arteries, the entire supply area covers parts of the striatum, the central corpus nuclei caudati and the internal capsule in between. The spatial insertion of a typical infarct then uncovers that the gray matter only is affected by hypoxic tissue changes, whereas the fibers of the internal capsule are completely spared (Fig. 6). Thus, size and shape of the infarcts not only depend on the mentioned conditions, such as vascular supply and possible substitution, but equally on tissue susceptibility. "Selective vulnerability", a phenomenon commonly associated with borderline and reversible hypoxic damage, might also therefore play some role in territorial infarction.

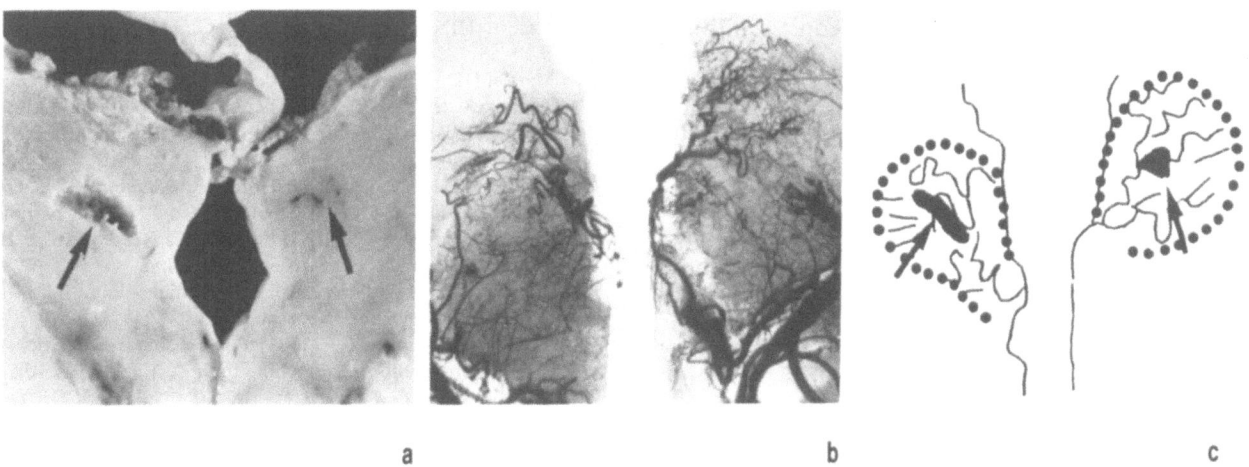

Fig. 5. (a) Bilateral almost symmetrical infarct within the territory of posterior thalamoperforating arteries. (b) Visualization of the branches of this arteries by selective angiography of both posterior cerebral arteries. (c) Schematic representation of the vascular supply area

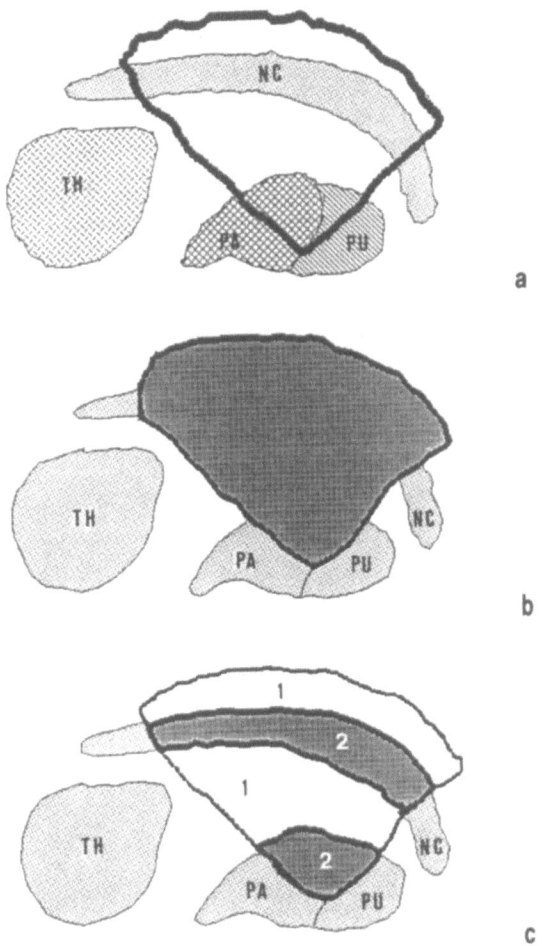

Fig. 6. (a) Picture of the area of supply of the striatal vessels (strong line) superimposed on anatomical structures: *NC* nucleus caudatus, *TH* thalamus, *PA* pallidum, *PU* putamen. (b) Complete infarction of this area (dark). (c) Selective infarction in the grey substance *2*, while the white matter *(1)* is unchanged

Differential diagnosis of cerebral lesions has to consider the tremendous variability of aspects presented by them. Spatial reconstruction, by imaging or computer, is essential for properly evaluating the three dimensional anatomical and pathological conditions detected by direct insight into the structures. In addition, spatial reconstruction may by better representation lead to clearer pathogenetic concepts.

Space Occupying Lesions, Mass Displacement and Conclusions

Neuroendoscopy up to now seems to have few indications within the intracranial and intraspinal cavities. The most frequent application is probably the approach of lesions confined to or in connection with the cerebrospinal fluid. Such lesions might be better diagnosed by their visible appearance and endoscopically guided stereotactic probes. This however presupposes familiarity with the anatomy and pathology of the ventricular system, the basal cisterns and the subarachnoid space. In these intracranial compartments determined sequelae of both local and general space occupying lesions take place. Starting from tumour, haemorrhage, suppuration or infarction, brain tissue is displaced by herniation into cisterns. Multiple consequences arise by a general increase in pressure within the intracranial space eventually surpassing blood pressure with fatal outcome. These events belong to the essential knowledge of clinical neurology and neurosurgery[19]. Size changes and displacements of spinal fluid compartments have to be taken into account, when endoscopic scouting is performed.

Possibly, wider indications will concern conditions within the brain parenchyma, especially cystic ones, as it is already the case with tumour cysts. Since brain tissue for a great many injuries has only a single response, namly colliquative necrosis, which ultimately leads to cyst formation there are many candidates for endoscopic procedures considered purely from the morphological angle. [To possibly approach this field, notions concerning the internal tissue make up and the steric composition of presumable compartments are necessery.] This presentation is meant to suggest exchange of views held in neuropathology and gained step by step in neuroendoscopy.

References

1. Benedum J (1988) Das Riechorgan und seine Erforschung bis S. Th. Soemmering. In: Mann G, Benedum J, Kümmel WF (eds) Soemmering-Forschungen III. Gehirn—Nerven—Seele Anatomie und Physiologie im Umfeld S.Th. Soemmerings. Akademie der Wissenschaften und der Literatur, Mainz, Gustav Fischer, Stuttgart
2. Broca P (1861) Sur le siège de la faculté du langage articulé. Bull Soc Anat (Paris) 2: 355
3. Dryander J (1537) Anatomiae hoc est, Corporis humani, dissectionis pars prior, Marburg
4. Henderson VW, Fich CB (1989) The neurobiology of Alzheimer's disease. J Neurosurg 70: 335–353
5. James CD, Carlbom E, Dumanski JP, Hansen M, Nordenskjold M, Collins VP, Cavenee WK (1988) Clonic genomic alterations in glioma malignancy stages. Cancer Res 48: 5546–5551
6. Kraepelin E (1983) Lebenserinnerungen. Springer, Berlin Heidelberg New York Tokyo
7. Mennel HD (1986) Bernhard von Gudden. Med Welt 37: 1449–53
8. Mennel HD (1988) Pathologie der intrakraniellen Tumoren als Grundlage der Prognose und Therapie. In: Graul EH, Pütter S, Loew D (eds) Medicenale XVIII. Das Gehirn und seine Erkrankungen (ii)
9. Mennel HD (1988) Geschwülste des zentralen und peripheren Nervensystems. In: Doerr W, Seifert G (eds) Spezielle pathologische Anatomie, Bd 13/III Pathologie des Nervensystems III. Springer, Berlin Heidelberg New York Tokyo
10. Mennel HD (1991) Grading of intracranial tumors following the WHO classification. Neurosurg Rev 14: 249–260
11. Mennel HD, Bingener J (1993) Postmortem examination of glioblastoma multiforme in whole brain sections. Tumordiagn Ther 14: 66–71
12. Möbius PJ (1925) Franz Josef Gall. Barth, Leipzig
13. Roßberg C, Boccalini P, Wagner H-J (1992) Über Infarkte im Versorgungsgebiet der Arteriae lenticulostriatae. Eine neuropathologische und postmortal-neuroradiologische Analyse. Klin Neuroradiol 2: 75–84
14. Roßberg C, Mennel HD (1986) Über ein- und doppelseitige Infarkte im Versorgungsgebiet der A. thalamoperforata posterior: Neuropathologische Befunde. Nervenarzt 57: 29–34
15. Spielmeyer W (1922) Histopathologie des Nervensystems. I. Allgemeiner Teil, Springer, Berlin
16. World Health Organization (1979) Histological typing of central nervous system tumors WHO, Geneva
17. Zülch KJ (1971) Atlas of the histology of brain tumors. Springer, Berlin Heidelberg New York
18. Zülch KJ (1977) Clinical ischemica: brain infarcts. In: Zülch KJ, Kaufmann W, Hossmann K-A, Hossmann V (eds) Brain and heart infarct. Springer, Berlin Heidelberg New York
19. Zülch KJ, Mennel HD, Zimmermann V (1974) Intracranial hypertension. In: Vinken PJ, Bruyn GW (eds) Tumors of the brain and skull. Handbook of clinical neurology, Vol 16. North Holland, Amsterdam/Elsevier, New York, pp 89–149

Correspondence: H.D. Mennel, M.D., Department of Neuropathology, Medical Center of Pathology, Philipps University Marburg, Baldingerstrasse 3, D-35043 Marburg, Federal Republic of Germany.

Acta Neurochir (1994) [Suppl] 61: 69–75

Endoscopic Diagnosis and Treatment of Para- and Intra-Ventricular Cystic Lesions

J. Caemaert, J. Abdullah, and **L. Calliauw**

Department of Neurosurgery, Hospital University Ghent, Belgium

Summary

Different cystic lesions can be located in or around the ventricular system, eventually causing hydrocephalus. Twenty-one patients are described where endoscopic intervention, mainly large fenestration towards the ventricular cavity, has been performed. This treatment can sometimes replace open surgery or extracranial shunting. Most rewarding are the arachnoid and ependymal intra- and paraventricular cysts. With careful and adequate endoscopic technique this procedure is safe and much less invasive than other methods described.

Keywords: Intra- and paraventricular cysts; cerebral endoscopy; laser.

Introduction

A surprisingly large number of para- and intraventricular cystic lesions of various nature are identified by CT and MRI imaging. Some of them cause hydrocephalus, others cause a variety of signs and symptoms due to their location and local distention. Endoscopic techniques are tending more and more to replace open surgery and extracranial drainage for these cases.

Twenty-one patients with such lesions are described in this paper from 1986 till April 1992. A cerebral endoscope has been designed by the first author and is described in a separate paper in this supplement, with details on techniques[2]. Four cases of suprasellar arachnoid cysts are described separately[3].

Material and Method

Twenty-one patients were treated for various lesions (summarized in Table 1). There were 14 males and 7 females and the ages varied between 2 weeks and 64 years.

1) Cystic Tumour

A 15 year old boy presented a postcomatose state after a short period of acute intracranial hypertension. CT scan revealed a large cystic lesion without obvious tumoural parts except a small sickle shaped contrast enhancement in the posterior part. The first diagnostic endoscopy showed an inner wall, very different from an arachnoid or ependymal cyst, with abnormal vascularisation mainly in the posterior region. Biopsy precipitated some venous bleeding which caused this procedure to be aborted. We considered the lesion to be an exceptional form of inoperable suprasellar haemangioblastoma. First attempt at treatment was the instillation of radioactive Yttrium-90 suspension. This was done without complications and a catheter with a Rickham reservoir was left in the cyst (Fig. 1) to deal with recurrent hypertension. The patient recovered very well but the Rickham reservoir had to be removed after 6 weeks because of skin necrosis over the dome of the reservoir. After this he was doing well until 7 months later when he again developed headache. CT showed an enlargement of the cyst (Fig. 2). Then we decided to perform a large fenestration towards the lateral ventricle (Fig. 3). After this the patient was perfectly well and is more fully employed as a bakers-man. It has to be said, that so far we have no definite pathological diagnosis in this case.

2) Colloid Cysts of the Third Ventricle

From our experiences colloid cysts tend to stick very tightly to the surrounding brain structures and choroid plexus at the foramen of Monro. Blind stereotactic puncture has been advocated, but in a large number of cases the contents are so viscous that it is not possible to evacuate a significant amount of mucous. Moreover a meer endoscopic inspection of the surroundings of the foramen of Monro makes clear that a blind puncture even stereotactically guided may be very dangerous for the fornix, the thalamostriate vein, the septal vein and the choroid plexus itself. Very often small vessels are running over the surface of the cysts and also the choroid plexus may be overlying the cyst wall partially. These vessels and plexus can be coagulated by a blunt laser fiber (0.6 mm) in the noncontact mode. Then the cyst wall can be opened by the cutting laser using conically shaped fibers in the contact mode. A large hole in the wall is made through which the viscous contents can be aspirated by a thin walled aspiration catheter with a broad lumen and by using the grasping forceps. A small grasping forceps can be introduced through the working channel or through one of the rinsing channels to grasp the capsule and to bring as much

Table 1. *Personal Cases*

	n
1. Paraventricular arachnoid and/or ependymal cysts	10
2. Suprasellar arachnoid cysts	4
3. Epidermoid cysts	3
4. Colloid cysts of the 3rd ventricle	3
5. Cystic tumour	1
Total number	21

vessels as possible in the reach of the coagulating laser fiber. Finally the small remnants of the empty cyst wall are removed. In this way very extensive removals can be performed by a unilateral frontal burrhole through one trajectory. In cases of asymmetrical hydrocephalus with a bulging septum pellucidum we prefer to make a large fenestration in the septum pellucidum rather than to make a second burrhole contralaterally. The patient can leave the hospital after one or two days.

By using this laser technique we expect recurrence cysts to be rare but if they occur they can be treated a second time endoscopically.

A very interesting paper on this subject is the work of Heikkinen (5) who reports a total removal of such a colloid cyst by endoscopy.

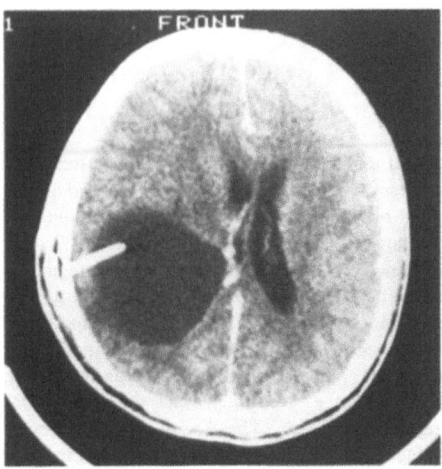

Fig. 1. Rickham reservoir in a cystic paraventricular tumour with Yttrium 90 treatment that failed

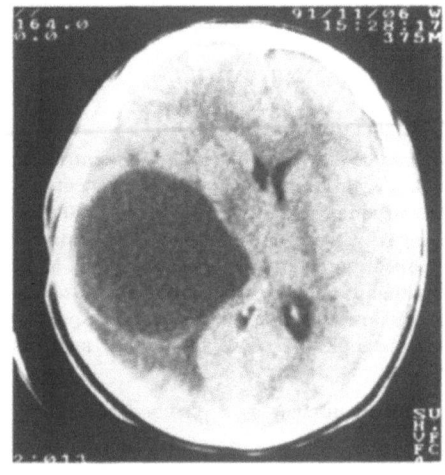

Fig. 2. Enlargment of the cyst of Fig. 1, 7 months after Yttrium instillation

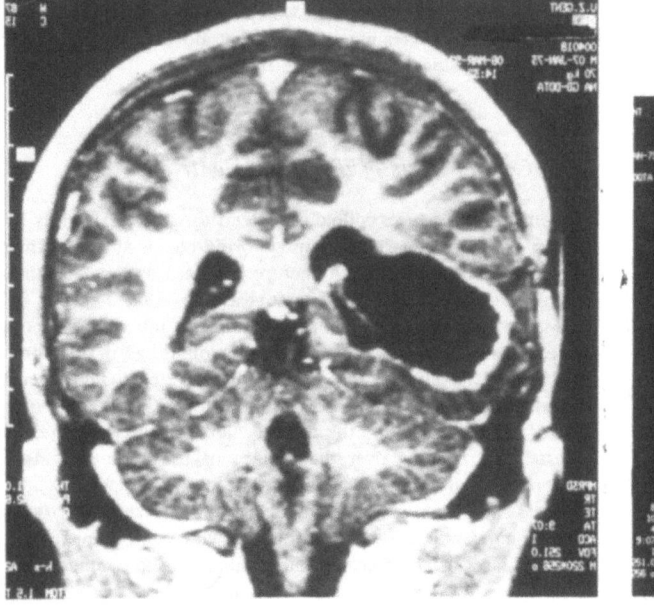

a

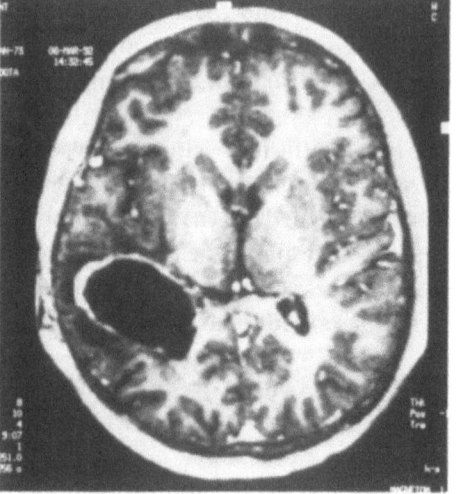

b

Fig. 3. (a) Coronal and (b) axial MRI 4 months after fenestration of tumoural cyst towards the lateral ventricle

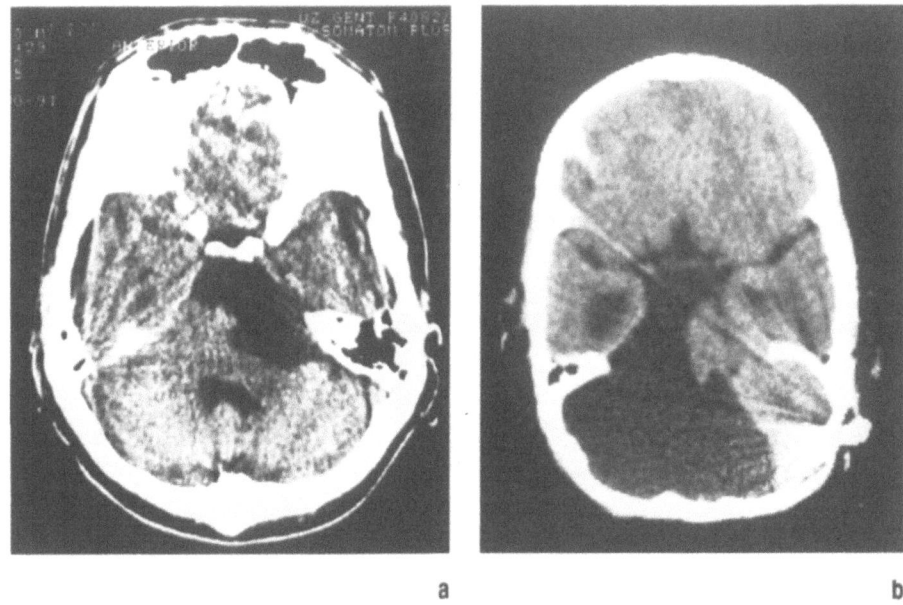

a   b

Fig. 4. (a) Epidermoid cyst versus (b) arachnoid cyst in posterior fossa with very similar attenuation patterns

3) Epidermoid Cysts

Although this lesion is very different from other intra-or paraventricular cysts, it is worth mentioning that they can be mistaken for CSF containing cysts, since their attenuation is very similar or identical to that of arachnoid or ependymal cysts (Fig. 4).

In these cases the endoscope is mainly a diagnostic tool since the typical contents are very viscous, so that aspiration is impossible even through a thick polyurethane catheter.

Nevertheless even open surgical treatment in the vast majority of cases only consists of an emptying of the caseous contents from within the capsule which moulds into every cleft and folding of the surrounding brain.

In future we will therefore treat cases by vaporization with the cutting laser, while aspirating at the same time. For this purpose a flexible optic element that fits into the working channel of our rigid endoscope is under development. On the other hand a definite danger in doing this is, that larger vessels crossing the epidermoid in one of its often numerous foldings, might be cut and cause uncontrollable bleeding. Personally we are very sceptical about optimistic reports concerning the possibility of coagulating bleeding vessels of more than one millimeter diameter. One reason is that the visibility in CSF is much reduced or even completely impaired by bleeding; another reason is that a ruptured vessel can retract out of reach. Moreover bleeding vessels larger than one millimeter are extremely difficult to deal with by endoscopic tools even in ideal circumstances of visibility and ability to reach them.

Figure 5 shows an epidermoid cyst extending from the pontocerebellar angle towards the temporal horn of the lateral ventricle. This cyst caused seizures and was mistaken for an arachnoid cyst. An attempt was made to perform a fenestration towards the ventricle. To our surprise, we found endoscopically, an epidermoid cyst filling the whole cavity, although a vermis agenesis at first sight made us believe, preoperatively that it was an arachnoid cyst. In a second step the patient was operated upon conventionally.

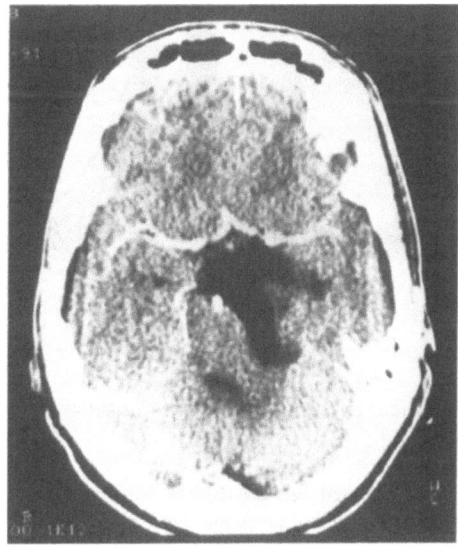

Fig. 5. Epidermoid cyst in pontocerebellar angle reaching up to the temporal horn of the lateral ventricle

4) Suprasellar Arachnoid Cysts

Four children were treated endoscopically. They are reported in detail in a separate paper[3]. The same technique is advised, as described subparagraph E. in this paper. None of these children needed a shunt after fenestration of the cyst into the lateral ventricles and the basal cisterns. There was no mortality and the morbidity was limited to one postoperative epileptic attack with full recovery.

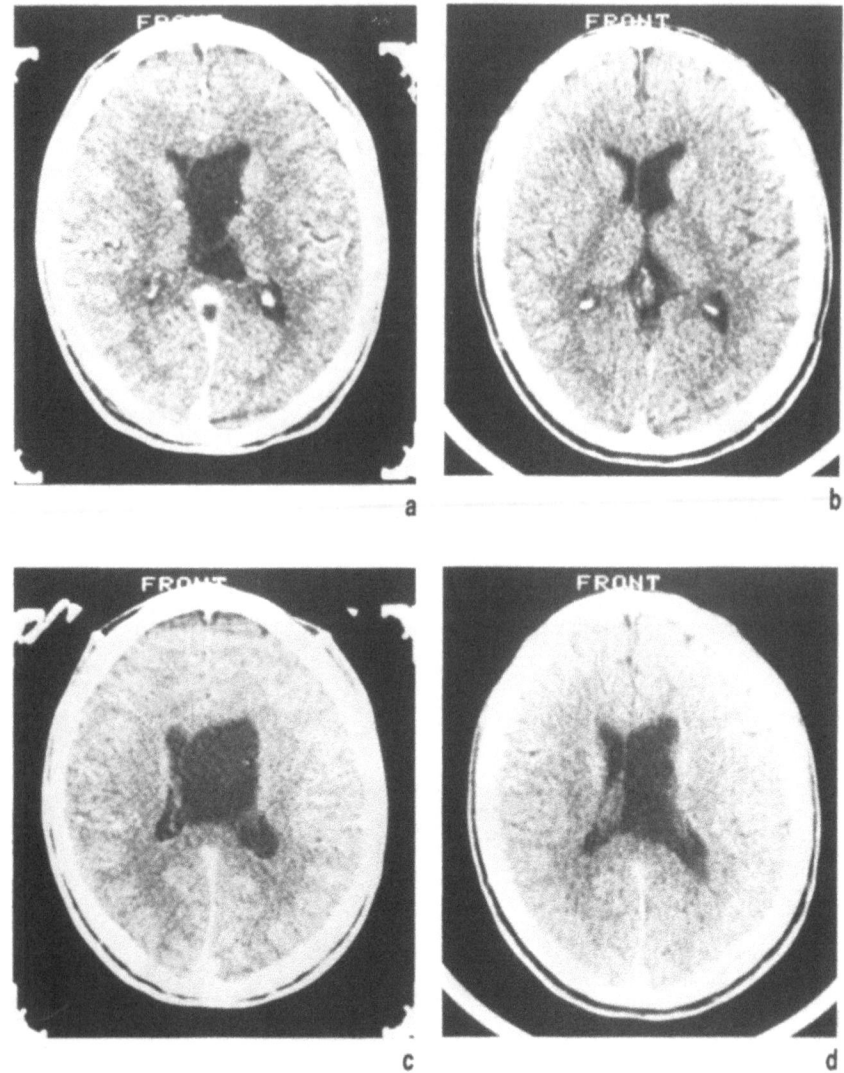

Fig. 6. (a) and (c) Intraventricular ependymal cyst before and 1 year after (b and d) endoscopic fenestration

5) Para- and Intraventricular Arachnoid and/or Ependymal Cysts

Ten cases have been treated between March '90 and April '92. These cysts are being recognized more and more by CT and MRI (Fig. 6). In the past, shunting procedures for hydrocephalus sometimes failed, because septations may divide the ventricles into separate compartments[7,8], so that drainage by insertion of a catheter was limited to one compartment (Fig. 7).

Other cysts do not produce hydrocephalus but are symptomatic due to their location or local distention. They cause most often epilepsy and headache. Owing to their periventricular topography the limbic system is sometimes disturbed with neuropsychological impairment.

Table 2 summarizes the cases with their symptoms and results.

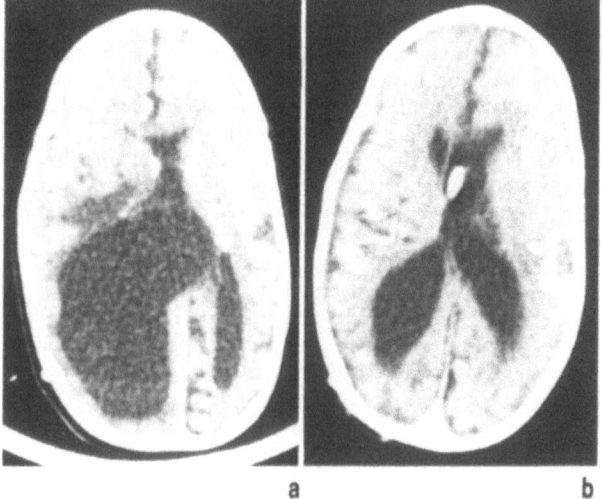

Fig. 7. (a) Occipital intraventricular cyst causing hydrocephalus by shift of aqueduct. This cyst is not drained by the right frontal catheter and persists until (b) fenestration has been performed

Table 2. *Ten Personal Cases with Para- and Intraventricular Arachnoid and/or Ependymal Cysts*

	Age	Sex	Location	Symptoms	Operation	Complications	Results	F.U.
1. VDGG	21y	m.	left lat. ventr.	headache concentrating difficulties	fenestration	—	good	25 m.
2. JM	3m	m.	interhemispheric posterior	hydrocephalus 1 month prior to endoscopic operation a ventriculo-peritoneal shunt inserted	fenestration	—	insufficient; 2nd larger fenestr. planned	20 m.
3. DVT	1y	m.	porencephalic	delayed milestones uncontrolled epilepsy	fenestration	—	good; epilepsy is well under control	10 m.
4. DPR	7m	m.	fossa posterior cyst	hydrocephalus ventriculo-peritoneal shunt 4/12 months prior to endoscopic operation	fenestration towards cisterna magna	—	good	10 m.
5. TB	1.5y	m.	porencephalic cyst left occipital horn	epilepsy	fenestration 1st trial abandoned due to bleeding 2nd trial uneventful	—	good	8 m
6. VGB	15y	m.	subependymal cyst right lat. ventr. trigonum	headache memory loss	fenestration	—	good	4 m.
7. DLJ	1y	m.	right occipital horn ependymal cyst	uncontrolled epilepsy hydrocephalus ventriculo-atrial shunt 1 year prior to endoscopic procedure	fenestration	—	good; epilepsy is well controlled	3 m.
8. LP	23y	m.	right occipital horn ependymal cyst	headache	fenestration	—	good	4 m.
9. VDM	1m	f.	interhemispheric posterior arachnoid cyst	hydrocephalus meningomyelocoele	fenestration	uneventful operation but postop. ARDS after 5 hours	+	
10. BC	45y	f.	right occipital horn ependymal cyst	headache	fenestration	—	good	2 m.

F.U. Follow-up.

Surgical Technique

Very important is the choice of approach to the cysts. Since our endoscope is a rigid one, all fenestrations that are planned have to be located in one straight line.

One may come first through the ventricular cavity, and from there penetrate towards the cyst cavity, or vice versa depending upon their spatial relationship.

Whether they should be approached stereotactically or free-hand, depends upon the size of the ventricles and the specific location of the lesion. In many cases stereotactic guidance is very useful to control the direction and depth of penetration in order to reach the target. Once on target, one can detach the endoscope from the stereotactic arc to have more freedom for delicate movements of the shaft. In many cases the hands of an experienced assistant are better than any supportive device for free hand endoscopy. The assistant can very quickly and delicately correct the direction and depth of the shaft at the request of the surgeon who handles the instruments.

Several techniques have been tried successively to perforate and fenestrate the membranes between cyst and ventricle[2,3]. Finally we prefer largely the laser for coagulation of blood vessels and cutting of a window in the cyst-walls (MBB Medilas 4060).

Coagulation of small vessels often abundantly crossing the membranes is very easily achieved by means of a blunt fiber (0,6 mm), in the non-contact mode using 20–40 W power.

Only the vessels need to be coagulated since coagulation of the whole trajectory that has to be cut, produces complete blanching of this track and, since this white surface reflects the laser light, much higher power for cutting is needed. So we can easily cut a "roundel" out of the cyst wall using the fibertome mode 2 (feed back control of the temperature of the tip to 700° C).

To prevent it from falling into the ventricle and causing aqueduct obstruction, the last fine attachment of this cut-off "roundel" is grasped with a forceps, avulsed and extracted.

Results and Complications

Of these ten patients seven were improved, with disappearance of pre-operative symptoms and without complications. One child had to undergo a second operation very recently. Treatment was carried out before we used the laser, and the hole made by perforation and enlargment by means of a Fogarty balloon catheter was too small and healed. It was very interesting to see that the first hole was not closed completely, and unless a residual opening of about 2,5 millimeter remained, the cyst wall was very bulging. During the second intervention we made an opening of 20 mm by 10 mm and removed the "roundel". A second smaller hole was made towards the trigone of the contralateral lateral ventricle.

The first attempt in case 5 had to be interrupted at an early stage because venous bleeding impaired vision to such an extent that rinsing was not sufficient to allow to carry on. The second trial one week later was uneventful and successful.

In case nine the fenestration by laser application was completely uneventful and after the endoscopic procedure, at the same operative session, its large meningomyelocoele was closed after 1 month of healing per secundam. The narcosis had been very difficult and took 4 hours, but intra-operatively there were no anaesthesiological problems. Five hours after the in-

tervention the child, who was already fully awake and without neurological symptoms (except for the lower limbs due to the meningomyelocoele), developed a very severe acute respiratory distress syndrome and died twelve hours later.

Discussion

Endoscopic fenestration of para- and intraventricular cysts is a valuable method that in most cases replaces an open surgical intervention or extracranial shunting procedure. Among other endoscopic procedures this is, in our experience the most frequent (21 cases of a total series of 45) and most rewarding intervention. Nevertheless the technique is fraught with specific problems. The choice of approach is of the utmost importance and determines the outcome of the intervention.

The greatest danger is intraoperative bleeding. If this occurs even if only by rupture of a small vessel, simple rinsing may be sufficient to restore clear vision and to stop the bleeding. Eventually bipolar coagulation or laser coagulation in the non-contact mode should be used.

In cases where bleeding is uncontrollable but not excessive, an external ventricular drainage for some days proves very helpful but was not necessary in our series.

We had no infection but systemic antibiotic prophylaxis was given in every case for 24 hours (amoxycilline trihydrate and potassium clavulanate).

With the laser the procedure is much easier and safer than with the initial method of bipolar coagulation and enlargment of the primary hole by repeated dilatations by an inflatable balloon catheter. The latter technique remains very useful when a hole has to be made in small areas near to large vessels (for example the basilar artery). In cases where the ventricular system is normal, drainage of the cyst can be expected by the normal CSF pathways. In cases where the cysts cause hydrocephalus (for example the four suprasellar arachnoid cysts), this may be relieved by fenestration, provided that the resorptive capacities over the convexity are normal. If this is not the case, the fenestration is nevertheless necessary prior to an extracranial shunting procedure.

References

1. Auer LM, Holzer P, Ascher PW, Heppner F (1988) Endoscopic neurosurgery. Acta Neurochir 90: 1–14

2. Caemaert J, Abdullah J, Calliauw L (1994) A multipurpose cerebral endoscope and reflections on technique and instrumentation in endoscopic neurosurgery. Acta Neurochir (Wien) [Suppl] 61: 49–53

3. Caemaert J, Abdullah H, Calliauw L (1992) Endoscopic treatment of suprasellar arachnoid cyts. Acta Neurochir (Wien) 119: 68–73

4. Gentry LR, Smoker WR, Turski PA, *et al* (1986) Suprasellar arachnoid cysts 1. CT recognition. AJNR 7: 79–86

5. Heikkinen ER (1986) Stereotactic neurosurgery: new aspects of an old method. Ann Clin Res 18 [Suppl 47]:

6. Leo JS, Pinto RS, Hulvat GF, *et al* (1979) Computed tomography of arachnoid cysts. Radiology 130: 675–680

7. Powers SK (1992) Fenestration of intraventricular cysts using a flexible, steerable endoscope. Acta Neurochir (Wien) [Suppl] 54: 42–46

8. Schultz P, Leeds NE (1973) Intraventricular septations complicating neonatal meningitis. J Neurosurg 38:620–626

9. Zamorano L, Chavantes C, Dujovny M, Malik G, Ausman J (1992) Stereotactic endoscopic interventions in cystic and intraventricular brain lesions. Acta Neurochir [Suppl] 54: 69–76

Correspondence: J. Caemaert, M.D., Department of Neurosurgery, University Hospital, B-9000 Ghent, Belgium.

Acta Neurochir (1994) [Suppl] 61: 76–78

Stereotactic Techniques for Colloid Cysts: Roles of Aspiration, Endoscopy, and Microsurgery

D. Kondziolka and **L.D. Lunsford**

Department of Neurological Surgery and the Specialized Neurosurgical Center, University of Pittsburgh
School of Medicine, Pittsburgh, PA, U.S.A.

Summary

Stereotactic surgical techniques were used to manage 25 consecutive patients with colloid cysts of the third ventricle. All patients had stereotactic aspiration as the initial procedure; it was successful in 13 (52%). In patients in whom aspiration failed, endoscopic visualization of the cyst and attempted removal, or microsurgical cyst resection via a transcortical approach was performed. Three elderly patients did not have a second procedure and were shunted. The computed tomographic appearance of a hypodense or isodense cyst indicates low-viscosity contents and predicts successful stereotactic aspiration. Stereotactic microsurgery is used in hyperdense cysts, or in those which cannot be aspirated.

Keywords: Colloid cyst; stereotactic surgery.

Introduction

Most colloid cysts require surgical treatment, but selection of the best management strategy for individual patients remains controversial. Although refinements in the transcallosal or transcortical approaches to the third ventricle have reduced the morbidity associated with these procedures, stereotactic guidance provides significant advantages over either technique.

For colloid cysts with low-viscosity contents, stereotactic aspiration has proved to be a successful alternative to open resection[3-6,10-12]. Cyst recurrence is rare. Stereotactic endoscopic removal allows the surgeon to visualize the cyst and adjacent structures, and to use a selection of instruments for cyst entry and evacuation. Stereotactic microsurgical resection via a transcortical approach provides a rapid and direct path to the cyst and minimizes brain dissection[15]. Standard microsurgical technique can then be used to remove cysts of varying sizes and consistency.

This paper will discuss our results using these techniques, and suggests guidelines for their use in individual patients.

Clinical Material

Over a ten-year period, 25 consecutive patients underwent computed tomography (CT)-guided stereotactic surgery for colloid cysts of the third ventricle. Patient age varied from 20 to 70 years. Four patients had bilateral ventriculoperitoneal shunts placed at the referring institution before undergoing stereotactic surgery. Our surgical technique has been reported previously[7].

Results

Table 1 details the results for each procedure (aspiration, endoscopic removal, or microsurgical resection). There was no morbidity or mortality after stereotactic intervention. All patients had aspiration as the initial surgical procedure; it was successful in 13 (52%). Success was defined as reduction in cyst volume to allow normalization of cerebrospinal fluid flow. Preaspiration cyst volume measurements were made from the CT scans[9]. The amount of colloid material aspirated was compared to the calculated preoperative volume. In the case of incomplete removal, a repeat aspiration was performed.

After failure of initial aspiration, 3 patients had attempted endoscopic removal. In one patient, 70% of the hard, thick cyst contents were removed using forceps and curettage. However, in the other two patients, endoscopic removal was unsuccessful due to poor visualization of a small cyst in one, and failure to remove solid contents in the other.

Eight patients had microsurgical resection after failure of initial aspiration. In two, a portion of the cyst wall adherent to the fornix was left in situ.

Discussion

Because of its simplicity and low risk, stereotactic aspiration of the cyst has been advocated[3-6,10-12].

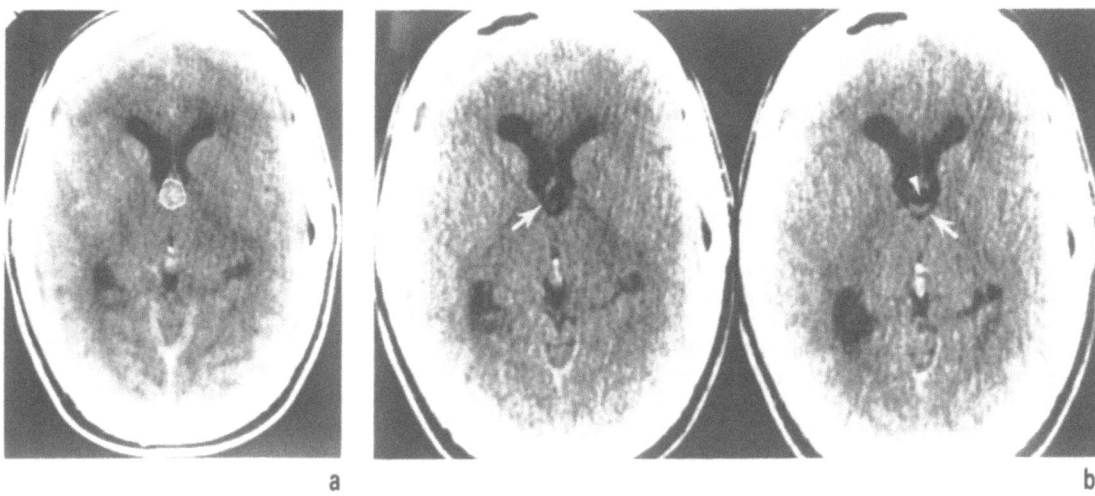

Fig. 1. (a) Pre-aspiration CT scan showing an isodense colloid cyst (outlined by the cursor function of the CT scanner). (b) Post-aspiration CT images show marked reduction in cyst size, air within the cyst (arrowhead), and patent foramina of Munro (arrows)

Table 1. *Results After Colloid Cyst Surgery*

	No residual cyst	Number of patients ——————— Small residual cyst	Over 30 % cyst volume remaining	Unsuccessful
Stereotactic aspiration	3	9	1	12
Stereotactic endoscopic removal	—	—	1	2
Microsurgery	6	2	—	—

Reports after aspiration have shown that total or subtotal removal of the colloid cyst contents does not require further treatment in most patients[4,6–8,10–12]. Because many cysts have hard, highly viscous contents, aspiration is not possible in all patients. We have reported those imaging features that successfully predict the viscosity of cyst contents, and thereby predict the usefulness of aspiration in specific patients[7]. The CT appearance of a hypodense or isodense cyst (as compared with brain on non-contrast studies) indicates a cyst with low-viscosity contents which can be aspirated. Aspiration failure in this setting can be due to deviation of the probe away from the cyst wall (in small cysts), or from an inability to puncture the cyst wall.

Some investigators have performed stereotactic endoscopic removal, although the number of procedures reported is small[2,4,7,12,13]. This procedure aims to provide a technique for removal of low or highly viscous cyst contents (by using the larger endoscopic cannula and available instruments) as well as the opportunity to visually inspect the cyst.

We have used stereotactic microsurgical resection via a transcortical approach to remove those cysts not treated successfully by aspiration. Stereotactic guidance permits precise placement of the corticotomy and provides a direct route to the cyst which minimizes brain dissection and retraction. It is not imperative that all the cyst capsule be removed, if doing so could cause injury to attached structures. Should a piece of cyst wall remain (as is usually the case after aspiration), the possibility of cyst reformation is extremely low[5].

Currently, our management strategy begins with careful interpretation of the CT and magnetic resonance images to determine whether or not the cyst can be aspirated. Asymptomatic colloid cysts 5 mm or less in diameter are usually followed conservatively with serial imaging, until cyst enlargement or clinical symptoms occur. For cysts larger than 5 mm, whether symptomatic or not, aspiration is recommended should the cyst be hypodense or isodense on CT. For patients with hyperdense cysts, we proceed to stereotactic craniotomy and microsurgery. We have not found that endoscopic removal provides significant advantages over a guided microsurgical approach.

References

1. Abernathey CD, David DH, Kelly PJ (1989) Treatment of colloid cysts of the third ventricle by stereotaxic microsurgical laser craniotomy. J Neurosurg 70: 195–200
2. Apuzzo MLJ (1988) Surgery of masses affecting the third ventricular chamber: technique and stragies. Clin Neurosurg 34: 499–522
3. Bosch DA, Rahn T, Backlund EO (1978) Treatment of colloid cysts of the third ventricle by stereotactic aspiration. Surg Neurol 9: 15–18
4. Caemart J, Calliauw L (1990) A note on the use of a modern endoscope. In Symon L et al (eds) Advances and technical standard in neurosurgery, Vol 17. Springer, Wien New York, pp 149–157
5. Donauer E, Moringlane JR, Ostertag CB (1986) Colloid cysts of the third ventricle. Open operative approach or stereotactic aspiration? Acta Neurochir (Wien) 83: 24–30
6. Gutierrez-Lara F, Patino R, Hakim S (1975) Treatment of tumors of the third ventricle: a new and simple technique. Surg Neurol 3: 323–325
7. Kondziolka D, Lunsford LD (1991) Stereotactic management of colloid cysts: factors predicting success. J Neurosurg 75: 45–51
8. Kondziolka D, Lunsford LD (1992) Factors predicting successful stereotactic aspiration of colloid cysts. Stereotact Funct Neurosurg 59: 135–138
9. Lunsford LD, Levine G, Gumerman LW (1985) Comparison of computerized tomographic and radionuclide methods in determining intracranial cystic tumor volumes. J Neurosurg 63: 740–744
10. Mohadjer M, Teshmar E, Mundinger F (1987) CT-stereotaxic drainage of colloid cysts in the foramen of Munro and the third ventricle. J Neurosurg 67: 220–223
11. Musolino A, Fosse S, Munari C (1989) Diagnosis and treatment of colloid cysts of the third ventricle by stereotactic drainage. Neurochirurgie 32: 294–299
12. Ostertag Ch B (1990) The stereotaxic endoscopic approach. In: Symon L et al (eds) Advances and technical standards in neurosurgery, Vol 17. Springer, Wien New York, pp 143–149
13. Powell MP, Torrens MJ, Thomson JLG (1983) Isodense colloid cysts of the third ventricle: a diagnosis and therapeutic problem resolved by ventriculoscopy. Neurosurgery 13: 234–237
14. Rivas JJ, Lobato RD (1985) CT-assisted stereotaxic aspiration of colloid cysts of the third ventricle. J Neurosurg 62: 238–243
15. Symon L, Pell M (1990) Surgical techniques in the management of colloid cysts of the third ventricle. In: Symon L et al (eds) Advances and technical standards in neurosurgery, Vol 17. Springer, Wien New York, pp 121–133
16. Yasargil MG, Sarioglu AC, Adamson TE, Roth P (1990) The interhemispheric-transcallosal approach. In: Symon L et al (eds) Advances and technical standard in neurosurgery, Vol 17. Springer, Wien New York, pp 133–143

Correspondence: Douglas Kondziolka, M.D., M.Sc., FRCS(C), Department of Neurological Surgery, Presbyterian University Hospital, Room 9402, 230 Lothrop Street, Pittsburgh, PA 15213, U.S.A.

Acta Neurochir (1994) [Suppl] 61: 79–83

Neuroendoscopic Third Ventriculostomy
A Practical Alternative to Extracranial Shunts in
Non-Communicating Hydrocephalus

R.F.C. Jones[1], B.C.T. Kwok[1,2], W.A. Stening[1,2], and M. Vonau[1,2]

[1] Department of Neurosurgery, Prince of Wales Children's Hospital and [2] Department of Neurosurgery,
Prince of Wales Hospital, Sydney, Australia

Summary

The outcomes in 103 patients who have undergone third ventriculostomy for non-communicating hydrocephalus at our institution form 1978–1994 have been analysed. The group has been sub-divided by age, cause of hydrocephalus and whether the third ventriculostomy was the initial definitive procedure or whether progression of their hydrocephalus had been arrested for a long time (usually years) by an extracranial shunt prior to third ventriculostomy. At the time of shunt malfunction (usually blockage) a third ventriculostomy was performed if the anatomy was, or could be made suitable for the safe performance of the procedure.

Third ventriculostomy under the age of 6 months was successful in only 8 of 25 patients. Seventeen patients in whom the onset of hydrocephalus was under the age of six months and the ventriculostomy was performed between 6 months and 18 years, 8 were successful. Sixteen of these had had previous long term shunts.

In 40 patients in whom the onset of hydrocephalus was over the age of 6 months and the ventriculostomy performed after the age of 19 years, 32 were successful. In 28 patients over the age of 20 years, 13 had previously been shunted and in 8 of these the procedure was successful. In 15 patients not previously shunted, 9 ventriculostomies were successful. Three failed, 2 died before evaluation could be done and one was lost to follow-up. There were no deaths caused by the procedure. Two patients suffered from a hemiparesis, (1 transient) 1 patient suffered mid-brain damage. There were 2 subdural effusions. Two patients had infections, 1 superficial and 1 a ventriculitis.

Keywords: Endoscopy; hydrocephalus; shunts; ventriculostomy.

Introduction

Endoscopic Third Ventriculostomy dates back to 1922[12]. Initially, the instruments used were bulky and the lighting and lens systems were poor. With the advent of the Hopkins lens system and fiberoptic illumination, miniaturisation was possible together with good visualisation of the anatomy achieved. The modern era of endoscopic third ventriculostomy was ushered in by Guiot[2] and has supplanted the radiologically guided stereotactic procedures. Localisation of the exact site of the endoscope by stereotactic, radiological (e.g. ventriculogram) ultrasonic or magnetic resonance imaging can be very useful at times.

Following the introduction of valve regulated extracranial shunts, there was initially thought to be no place for third ventriculostomy. When it became apparent that many complications occurred following placement of these implants, further attention was paid to third ventriculostomy[4], only to slacken as the results of extracranial shunting improved in the late 1970's and early 1980's.

We started using the antisiphon device routinely in 1978[9] and as a consequence acquired a significant number of patients so treated with relatively generously sized ventricles even when their intraventricular pressures were normal as have others[1].

A number of these patients still developed complications such as shunt blockage and with the shunt blockage the already generous ventricles became larger. Stimulated by the work of Vries[19], we decided to investigate the feasibility of third ventriculostomy as an alternative to shunt revision.

Initially, we selected only patients with very attenuated third ventricular floors bulging down into the interpreduncular cistern. Recently, we have relaxed our prerequisites and now do not demand this degree of attenuation. However, we still prefer a wide third ventricle ie. a few millimeters wider than the diameter of the ventricular scope, to permit the safe introduction of the endoscope. This can be achieved by waiting for the raised intracranial pressure to distend the third ventricle or in some patients with a functioning shunt to increase the pressure of the shunt. An alternative is to use an extraventricular drain and adjust the height and hence the intracranial pressure to achieve the same result.

Our investigations, instruments and techniques have been reported in detail elsewhere[7,10,18]. We do not attempt to assess the patency or otherwise of the csf absorptive pathways by isotope or infusion tests as we do not believe these are adequate[5,14].

In summary, structural studies are necessary to assess the site of the blockage in hydrocephalus. Computerised tomographic examinations have been the mainstay in the past but magnetic resonance examination is our preferred technique currently. The vast majority of our patients have been treated with a rigid endoscope, initially a needlescope and recently a Storz arthroscope. Recently we have also used a Codman flexible ventriculoscope.

There are many techniques available for third ventriculostomy[3,4,5,11,13,15-19]. We have been using the blunt puncture technique of Vries in the majority of cases but in a significant number the initial opening is made with a probe or a pair of forceps and then dilated to accommodate the ventriculoscope itself. This ensures that the opening is at least the size of the ventriculoscope. As an alternative to forceps, a Forgarty catheter can be used blowing up the balloon of the catheter when it is in the interpeduncular cistern and then drawing it into the third ventricle[15]. We have not used diathermy or lasers for fenestration and have not used stents[5]. We use a miniature camera and operate from the television screen.

Results (Table 1)

Group A

Patients who developed non-communicating hydrocephalus and who underwent third ventriculostomy under 6 months of age.

1. Fifteen patients without associated myelomeningocoele.
   Two patients had been previously shunted and in one the third ventriculostomy was a success and in one a failure. In the 13 patients with no previously placed shunt, 6 were successful. One of these was only successful on the second attempt and in another the ventriculostomy resealed 42 months later. In one patient, the initial progress seemed good but a recurrence of his bleeding tendency led to subdural haematoma and he later died of unknown cause in a hospital in New Zealand. He suffered from osteopetrosis treated by a bone marrow transplant. 6 of this group were failures.
2. Ten patients with associated myelomeningocoele.
   One patient had a previous shunt and in this patient the procedure was a failure. There were 2 successful outcomes, 5 failures

and 2 patients in whom improvement seemed to occur for 2 months only but who required a shunt.

Group B

Patients who developed hydrocephalus under the age of 6 months and in whom the third ventriculostomy was performed from 6 months to 18 years of age.

1. Without associated myelomeningocoele. 17 patients comprised this group, 15 had aqueduct stenosis, 1 was associated with an encephalocoele and one had a blockage secondary to cryptococcal meningitis.
   Sixteen had previously been shunted and 8 of these had a successful result. Of the 9 failures, 1 had previously been shunted for communicating hydrocephalus, 1 had a family history of hydrocephalus and 1 did not return for removal of his shunt.
   One of the patients was successfully controlled for 60 months and then developed headaches. She underwent intracranial pressure monitoring and showed a few short rises in intracranial pressure. In view of her previous history of visual obscuration with shunt malfunction, a shunt was placed. This did not improve her headaches and in retrospect, was a mistake.
2. With associated myelomeningocoele 12 had previously been shunted. In one patient a shunt was placed to control a csf leak in the neonatal period. She was referred to us as csf continued to leak from the abdominal wound at the age of 1 year. The shunt was removed. In 11 of these (including the girl with the csf leak) the procedure was successful. One remains shunt dependant and in the second a shunt was placed 9 months after third ventriculostomy as the parents insisted that the head size be made stationary.

Group C

Patients who developed hydrocephalus over the age of 6 months and who underwent third ventriculostomy up to the age of 19 years.

1. Sixteen patients had an obstruction at the aqueduct which was due to a tectal tumour in 4. In another 3 there was a block at the exit foramina of the fourth ventricle making a total of 19 patients in all. Five had previously been shunted and in 3 the third ventriculostomy was successful although 1 required a second attempt. One who was referred to us presented with massive post-shunting subdural collections (no ASD used) was improved only.
   Of the 14 patients who had no previous extracranial shunts, 12 were successful. In one of the failures, the ventriculostomy was patent at ventriculoscopy. The other patient's hydrocephalus was due to cryptococcal meningitis.
2. In 10 patients the blockage was due to a tumour originating elsewhere than the tectum. In 3 of these patients a third ventriculostomy was performed successfully to relieve the raised intracranial pressure prior to tumour removal. In another 2 the third ventriculostomy was successfully used in palliation—one of these had previously been shunted extracranially prior to subtotal resection. Both died from tumour progression with a functioning third ventriculostomy.
   In 5 patients the third ventriculostomy was performed when the extracranial shunt blocked and it was successful in 4 patients. One patient required a repeat third ventriculostomy after 32 months and this had been successful.

Table 1. *Ouctome After Third Ventriculostomy*

	Complete series			Following previous extracranial shunt		
	Number	Success	Failure	Number	Success	Failure
Onset under 6 months Ventriculostomy under 6 months	25	9	16	3	1	2
Onset under 6 months Ventriculostomy 6 months–18 years	29	19	10	28	19	9
Onset over 6 months Ventriculostomy 6 months to 19 years	26[a]	21	5	10	7	3
20 years and over at presentation	23	14	9	8	5	3
Totals	103	63	40	49	32	17

[a] Excludes 3 patients who had a third ventriculostomy successfully performed shortly before tumour resection.
Those patients lost to follow up, in whom the results were uncertain or who only improved have been classified as failures in the interest of simplicity.

Group D

Patients 20 years or over at presentation. Twenty-three patients.

These were patients with aqueduct blockage of undetermined aetiology except for one patient with a pineal tumour, one with cerebellar metastases and one due to thrombosis of the basilar artery presenting with a dilated, hypertensive ventricular system.

Eight patients had previously had extracranial shunts placed and in these a third ventriculostomy was successful in 5—though one required a second attempt before the raised intracranial pressure was relieved.

Of the remaining 15 patients there were 9 successful outcomes and 6 failures. Two of the successful ventriculostomies resealed, one after 5 months and one after 38 months. Two patients died soon after operation, one of a basilar artery thrombosis and one probably due to pneumonia. One was lost to follow up.

Discussion

We have modified our previous categories moving the initial age division down to 6 months from 2 years[6,10,18]. This was done as it seemed that the infants over 6 months of age were more comparable with the older than younger age groups.

In addition we have subdivided each age group into one in which these was no shunt controlling the progression of the patients hydrocephalus prior to third ventriculostomy and a group in whom a shunt had been placed some years before third ventriculostomy.

In no instance was the shunt placed 6 weeks or so before the third ventriculostomy as recommended by Sayers[16].

All the shunts inserted were Heyer Schulte of various pressure ranges with associated Antisiphon Device (ASD). These shunts have been routinely used in our unit since 1978[9]. Employment of this system has resulted in an increased number of patients presenting with ventricles of sufficient size to permit the safe performance of a third ventriculostomy. It may have contributed to improved function of the surface absorptive mechanism resulting in a higher success rate following successful third ventricular fenestration. This however is speculative and it will be necessary to await the publication of other series in which third ventriculostomy has been performed after the long term control of hydrocephalus by other shunting systems. Other series of third ventriculostomy following long term shunt have been discouraging[4,5].

The results in the group aged <6 months show a poor success rate i.e., high incidence of shunts placed shortly after third ventriculostomy. As only 1 patient in this group had a preoperative shunt one cannot draw any conclusions from the preshunted group.

The procedure can be used to simplify treatment of shunt infections resistant to conventional treatment[8,11].

However, it is noteworthy that some unexpected success was encountered—one child was premature (mass 1200G). A third ventriculostomy was performed as the child had necrotizing enterocolitis with several wound infections caused by staphyloccocus aureus. It was thought a shunt stood a high chance of becoming colonised. It was gratifying to achieve a successful outcome.

The other groups shows a high incidence of success for the procedure with no difference apparent between the groups who were initially shunted and those who were not. This is contrast to the experience of other authors[4,5]. Similarly, in contrast to others, we have had a successful outcome in 2 of 3 patients who had hydrocephalus secondary to meningitis[5,13,15].

A number of these patients did not have any other structural study than a CT scan. Reliance was placed in those on the relative size of the lateral third and fourth ventricles. It is possible that some of these may in fact have communicating types of hydrocephalus and these would have contributed to the bad results.

Two patients were noted to have shunts that have been assessed as functioning after third ventriculostomy. These two clinically were thought to have malfunctioning shunts at the time but as the shunts have not been investigated by shunt function studies since, they have categorised as failures.

Occlusion of the third ventriculostomy has been recorded on 6 occasions. That is lower than reported by other authors[13]. In 3 patients a second attempt at third ventriculostomy has been successful. Two of these were done shortly after the initial procedure and one was done 32 months after the initial third ventriculostomy. In one patient, bleeding was encountered at attempted second ventriculostomy and this procedure was not successful.

Although we have had no deaths due to the ventriculostomy, we have had serious complications.

Two patients have suffered hemiparesis and in one the hemiparesis was permanent. Another patient has had some midbrain damage.

One patient has had to have a subdural collection treated and two have had infections, one a minor skin infection and the other a ventriculitis which responded quickly to treatment.

Detailed knowledge of the endoscopic anatomy and insistence on a clear view of the structures prior to any attempt at ventriculostomy is essential in the prevention of neurologic complications. Illustrations of the anatomy have been published[10,15] or are in press[6].

Assessment of the success or otherwise of a third ventriculostomy is difficult. It usually takes at least four or five days for intracranial pressure to settle down even in favourable cases such as acquired aqueduct stenosis. In patients who have an earlier onset of non-communicating hydrocephalus, the period can be longer as in the patient illustrated in Fig. 1. This patient improved very considerably after third ventriculostomy. The intracranial pressure became normal and the ventricular dilatation stabilised. However, the head circumference continued to increase at a rate greater than normal for some six weeks before finally settling down to run in parallel with the 98th centile.

To be categorised as successful, we feel that the

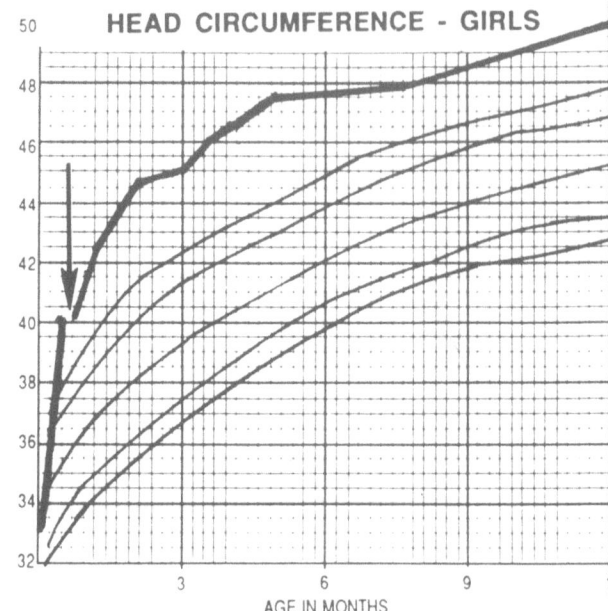

Fig. 1. This head circumference chart demonstrates the rapid increase in head circumference prior to third ventriculostomy (arrow). Thereafter, the child's fontanelle remained soft and the lateral ventricular ratio remained stable. The head circumference increased progressively, slowed over 5–6 weeks and then progressed parallel to the 98th centile. We interpret this as indicating that the subarachnoid pathways too, much longer than the usual 4–5 days to become functional. This is a further reason for not using isotope studies[5] or infusion tests[14] in selection of patients for third ventriculostomy

patient's intracranial pressure should be within normal limits. We do not insist that the ventricular size become normal and indeed it is unusual for it to do so. The assessment of the intracranial pressure is usually done clinically, supplemented at times with direct measurement or indirect measurement such as fontanelle tension or the tension of an artificial fontanelle. In addition, we see decreased ventricular size in the vast majority of patients often accompanied by increased fluid over the surface of the brain.

Magnetic resonance examination has proved very useful in the selection of patients for operation as one can visualise attenuation of the third ventricular floor preoperatively and hence, can exclude those patients in whom the third ventricular floor is unsuitable for the procedure.

It is, however, still not possible to assess which patients will develop suitable subarachnoid pathways. We have been able to get successful results in patients who have previously been treated for acute meningitis even in the neonatal period. However, we have understandably had some poor results in patients who have had previously communicating hydrocephalus or chronic meningitis. It would seem reasonable to exclude these patients, as the subarachnoid pathways would be occluded in a high percentage.

Conclusion

We believe that endeavour to improve the success rate by selection of patients should not be pursued too vigorously as some patients who might benefit may be excluded. However, chronic meningitis and previously communicating hydrocephalus with later acquired aqueduct stenosis seem to us to be contraindications as an occlusion of the subarachnoid pathways has already occurred.

Third ventriculostomy is now a proven successful method of controlling noncommunicating hydrocephalus. In our hands it is successful also in those patients who have previously had the progression of their hydrocephalus controlled by long term shunt. This has occurred even in some patients in whom meningitis has been associated with the onset of hydrocephalus.

Acknowledgments

We wish to thank the Department of Medical Illustrations, University of New South Wales for the diagram.

Especially, we wish to thank Cristina Kew for her assistance in preparation of the manuscript.

References

1. Gruber R, Jenny P, Herzog B (1984) Experiences with the antisiphon device (ASD) in shunt therapy of pediatric hydrocephalus. J Neurosurg 61: 156–162
2. Guiot G (1973) Ventriculocisternostomy for stenosis of the aqueduct of Sylvius. Puncture of the floor of the third ventricle with a leucotome under television control (with 6 figures). Acta Neurochir 28: 264–289
3. Hirsch JF, Hirsch E, Sainte-Rose C et al (1986) Stenosis of the aqueduct of Sylvius. Monogr Neurolog Sci 30: 86–92
4. Hoffman HJ, Harwood-Nash D, Gilday DL (1980) Percutaneous third ventriculoscopy in the management of non-communicating hydrocephalus. Neurosurgery 7: 313–321
5. Jaksche H, Loew F (1986) Burrhole third ventriculo-cisternostomy. An unpopular but effective procedure for treatment of certain forms of occlusive hydrocephalus. Acta Neurochir (Wien) 79: 48–51
6. Jones RFC, Kwok BCT, Stening WAS, Vonau M (1994) The current status of endoscopic third ventriculostomy in the management of non-communicating hydrocephalus: Minimally Invasive Neurosurgery 1. 37: 28–36
7. Jones RFC, Stening WAS, Bryden M (1990) Endoscopic third ventriculostomy. Neurosurgery 26: 86–92
8. Jones RFC, Stening WAS, Kwok BCT, Sands TM (1993) Third ventriculostomy for shunt infections in children. Neurosurgery 32: 855–860
9. Jones RFC, Teo C, Currie B, Kwok BCT, Nayanar VV (1991) The antisiphon device—its value in preventing excessive drainage. In: Matsumoto S, Tamaki N (eds) Hydrocephalus, pathogenesis and management. Springer, Berlin Heidelberg New York Tokyo, pp 383–390
10. Jones RFC, Teo C, Stening WAS, Kwok BCT (1992) Neuroendoscopic third ventriculostomy. In: Manwaring KH, Crone KR (eds) Neuroendoscopy, Vol 1. Liebert, New York, pp 63–78
11. Kelly P, Goerss S, Kall B, Kispert D (1986) Computed tomography-based stereotactic third ventriculostomy. Technical note. Neurosurgery 18: 791–794
12. Mixter WJ (1923) Ventriculoscopy and puncture of the floor of the third ventricle. Boston Med Surg J 188: 277–278
13. Pierre-Kahn A, Renier D, Bombois B, Askienay S, Moreau R, Hirsch JF (1975) Place de la ventriculo-cisternostomie dans le traitement des hydrocephalies non communicants. Neurochirurgie 12: 557–569
14. Pudenz RH, Foltz EL (1991) Hydrocephalus: overdrainage by ventricular shunts. Surg Neurol 35: 200–212
15. Sainte Rose C (1992) Third ventriculostomy. In: Manwaring KH, Crone KR (eds) Neuroendoscopy, Vol I. Liebert, New York, pp 47–62
16. Sayers MP, Kosnik EJ (1976) Percutaneous third ventriculostomy: experience and technique. Childs Brain 2: 24–30
17. Scarf JE (1966) Third Ventriculostomy by puncture of the lamina terminalis and floor of the third ventricle. J Neurosurg 24: 935–943
18. Teo C, Jones RFC, Stening WA, Kwok BCT (1991) Neuroendoscopic third ventriculostomy. In: Matsumoto S, Tamaki N (eds) Hydrocephalus, pathogenesis and management. Springer, Berlin Heidelberg New York Tokyo, pp 680–691
19. Vries J (1978) An endoscopic technique for third ventriculostomy. Surg Neurol 9: 165–168

Correspondence: Robert Jones, FC, FRACS, FRCS (Eng), Prince of Wales Children's Hospital, High Street, Randwick, NSW 2031, Australia.

Acta Neurochir (1994) [Suppl] 61: 84–91

Endoscopic Surgery of Traumatic Intracranial Haemorrhages

V.B. Karakhan and **A.A. Khodnevich**

Department of Neurology and Neurosurgery, Moscow Medical Stomatological Institute, Moscow, Russia

Summary

The results of endofiberscopic removal of traumatic intracranial haematomas and hygromas in 180 patients are analysed. Peculiarities of the surgical techniques using flexiscopes and original devices in epidural, subdural, intracerebral, intraventricular haemorrhages of various consistencies, size and location are reported.

A technique of the trephination access and delayed cranioplasty for endoscopic removal of extensive subdural and intracerebral haematomas is presented.

Indications, contra-indications for endoscopic haematoma surgery, advantages, disadvantages, failures of the techniques are discussed.

Keywords: Brain injury; surgery; cranioplasty; endoscopy; epidural haematoma; subdural haematoma; intraventricular haemorrhage.

Introduction

The main principle and aim of endoscopic haematoma neurosurgery is the removal of a great amount of blood by spilling as little as possible blood. For realization of this principle it is necessary to provide multilevel endoscopic stereotopography, adequate surgical technique and tools for precise actions in conditions of a wide spread of blood about and restricted visual field.

The surfaces of subdural and ventricular cavities posses a very complicated shape. Therefore one should prefer flexible endoscopes with bending tip for safe parabolic tube insertion, precise space-orientation and endoscopic navigation.

Our data on endoscopic subdural stereotopography and our clinical experience on endoscopic neurosurgery of traumatic intracranial haemorrhages were published in 1988[3] and 1994 for the first time. Now we summarize the results based on clinical experience of 233 endocranioscopic operations in neurotrauma.

Table 1. *Distribution of Patients with Traumatic Intracranial Haemorrhages Operated on Endoscopically*

Forms of haemorrhages	Severity of state			Volume of haematomas (ml)				Combined trauma
	Mode-rate	severe	critic	50	51–100	101–150	151–200	
Epidural (21)								
Acute (5)	—	5	—	—	1	3	1	—
Subacute (12)	8	4	—	—	12	—	—	—
Chronic (4)	4	—	—	2	2	—	—	—
Subdural (122)								
Acute (26)	1	14	11	—	16	9	1	3
Subacute (47)	17	25	5	—	23	11	7	—
Chronic (49)	30	16	3	1	19	20	9	—
Subdural hygromas (6)	—	2	4	1	4	1	—	2
Intracerebral (8)	4	4	—	2	6	—	—	—
Intraventricular (3)	—	1	2	3	—	—	—	1
Multiple[a] (20)	4	11	5	—	4	11	5	7
Total 180	68	82	30	9	87	61	23	13
		62.2%						

Male/female—152/28; Age—16–93 (mean —44.6).

[a] Bilateral (16): epidural-intracerebral; subdural-subdural (haematomas and hygromas); subdural-intracerebral; intracerebral-intracerebral. Unilateral (4): epidural + subdural; subdural + intracerebral.

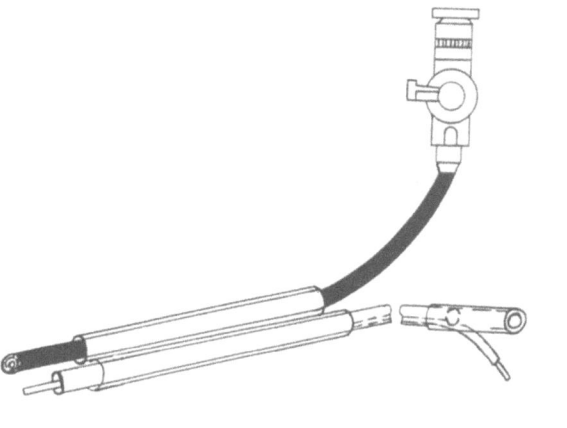

a

b

Fig. 1. (a, b) The sketch of the device used for removal of large clotted extracerebral haematomas including flexiscope, wide-channel draining system and uniting nozzle. (b) Endoscopic dissector

This allows one to formulate the indications for endoscopic removal of various kinds of intracranial haemorrhages.

Patients and Methods

208 patients were operated on endoscopically: 180—endoscopic surgery of traumatic intracranial haemorrhages; 28—diagnostic endoscopic exploration through a small cranial opening, bilateral in 4 cases. Patients of the last group were operated on in extremely serious condition with primary brain stem lesions in addition to the intracranial haematomas. This group is not analysed in the present paper.

Distribution of clinical cases is presented in Table 1. General methodology of neuroflexiscopy, techniques and equipment for endoscopic removal of intracranial haemorrhages have been reported in our previous paper[4]. For evaluation of large clotted haematomas the co-axial wide-channel nozzle was used (Fig. 1). This ensures a free axial shift of all components, insertion of endoscopic forceps (and) or original endoscopic "finger"-dissector[4] improving instillation-aspiration procedures for removal of clots.

A method of access to the subdural space and delayed cranioplasty without the second operation has been worked out and applied. Having made the linear scalp incision (7–8 cm), coronal trephination with bone disc removal is performed. The dura is opened in a manner shown in (Fig. 2a). The dural opening ensures subdural insertion of endofiberscope with co-axial draining system (the diameter of each being up to 6–7 mm) through the cranial opening of 20–25 mm in diameter.

After endoscopic removal of the haematoma a wide drain is inserted and passed through the dural incision. The other cuts of the dura are closed (Fig. 2b). Then the scalp area near the cranial opening is separated from pericranium, the bone disc is inserted on its outer surface under the scalp (Fig. 2b) and approximated to the trephine well. To anchor it in position a stitch is passed through a drill-hole in the bone disc and passed through the scalp.

To prevent damage to bridging pial vessels, drain insertion into the subdural space is performed at some distance from the trephination opening under endofiberscopic control. This helps to detect the number and location of these vessels (see Fig. 3).

The use of an optically translucent draining tube and the pericranial placement of the bone disc about the cranial opening for a distance not less than its diameter allows one to make: 1) a direct visual assessment of the drained cavity by insertion of the thin

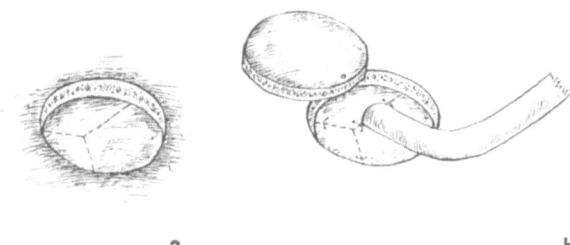

a   b

Fig. 2. Minimally invasive approach (a) and delayed nonsurgical cranioplasty technique (b) for endoscopic surgery of the intracranial haemorrhages. (a) Y-shaped incision of the dura; one long cut for insertion of coaxial nozzle or wide-channel drain; (b) position of bone disc and drain after clot removal and dural closure. Disc with drill hole

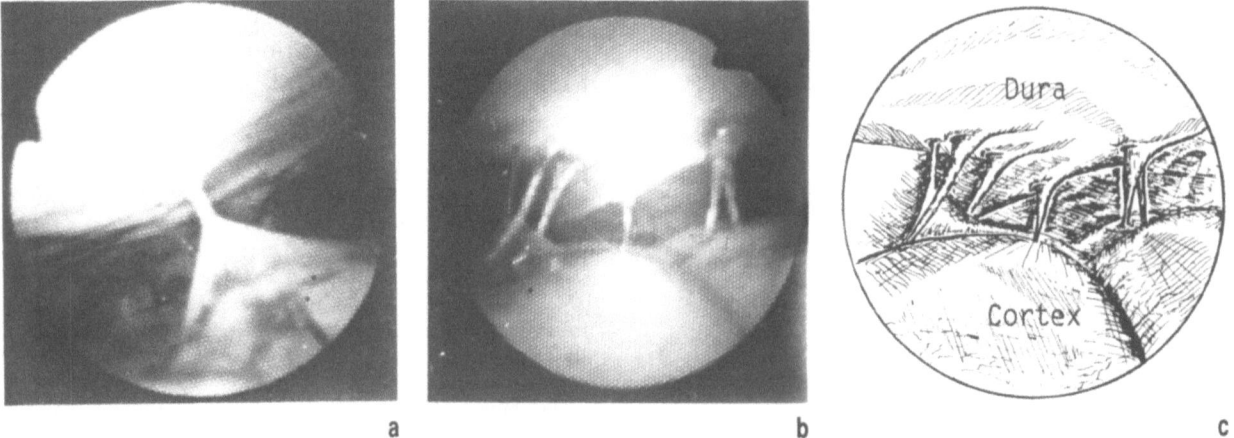

a   b   c

Fig. 3. Endoscopic subdural images of bridging artery (a) and veins (b) and sketch (c) retracting the cortical surface by superficial vessels and covering arachnoid

flexible fiberscope through the drainage tube (ENF-P "Olympus", 3.7 mm of tube diameter with high optic resolution—Fig. 3), and 2) ultrasonic sector scanning through the scalp for an evaluation of ventricular system displacement.

After the removal of the draining tube the bone disc is manoeuvred into the cranial opening and tying the ligature.

Results

Direct instillation-aspiration techniques using co-axial nozzle allows one to increase significantly the quality of endoscopic observation of blood-filled cavities and visualization of structures, especially bridging arteries and veins (Fig. 3). This ensures the reliable clearance of intracranial spaces of blood and the prevention of vessel/brain damage by the endoscopic tip or endomicroinstruments.

Peculiarities of Endoscopic Surgical Technique in Various Forms of Intracranial Haemorrhage

Epidural haematomas. Endofiberscopic technique ensures the evacuation in toto of large epidural haematomas of typical parieto-temporal (Fig. 4) as well as fronto-polar (Fig. 5) or occipital-posterior fossa locations (Fig. 6). The surface of the clots in contact with the inner table of the skull may exceed the plane of the craniectomy 8 to 13 times. In occipito-posterior fossa haematomas, depending on the haematoma extension, burr holes are placed near the fracture line above and below the transverse venous sinus (Fig. 7). This allows control of any bleeding from the sinus after the subdural blood has been evacuated, without enlargement of the trepanation.

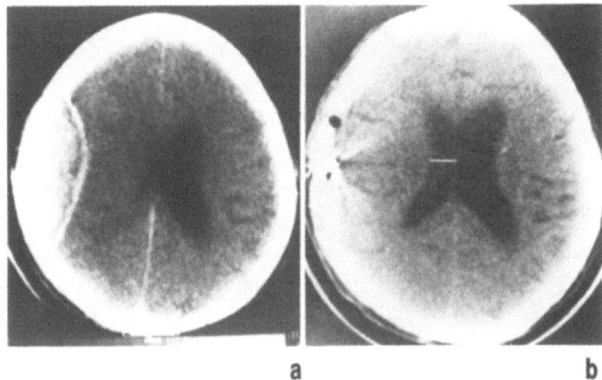

a   b

Fig. 4. CT before (a) and after (b) endoscopic evacuation of large epidural haematoma of typical localization

For prevention of new blood accumulation by oozing into the evacuated haematoma cavity the dura is made to stick to the inner table by use of fibrin glue and temporary application of pressure from the subdural side.

In acute forms (9, including 4 multiple haematomas) bleeding from branches of meningeal arteries was controlled by high-frequency coagulation beyond the cranial opening via the endoscope's instrumental canal. For amelioration of an acute tentorial pressure cone associated with severe midbrain syndrome a selective brief hydrocompression of the herniated uncus by endoscopic side-expanding microballoon[4] was used with reassuring preliminary results.

Subdural haematomas. Clotted and mixed types were completely removed endoscopically using aspiration—directed instillation technique with the help

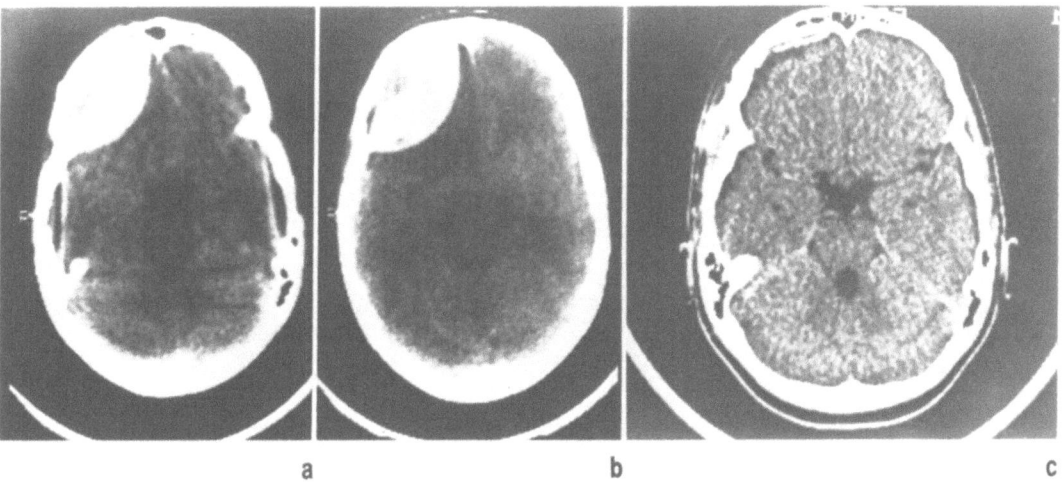

Fig. 5. Preoperative (a, b) and postoperative (c) CT in case of endoscopic removal of fronto-polar epidural haematoma without glue adhesion of separated dura to bone (free zone between dura and inner cranial surface (c))

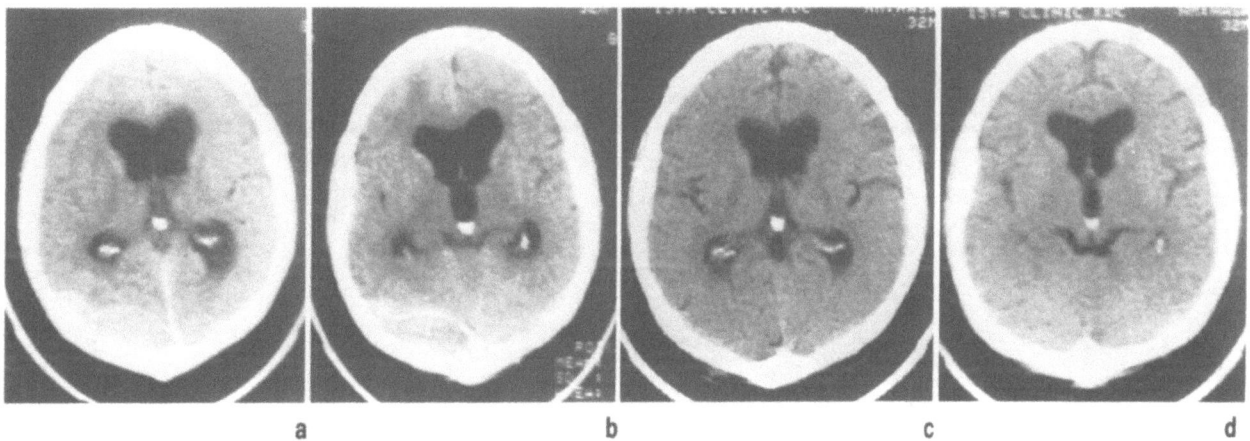

Fig. 6. Serial CT before (a, b) and after (c, d) endoscopic removal of epidural occipital-posterior fossa haematoma using glue attachment of separated dura

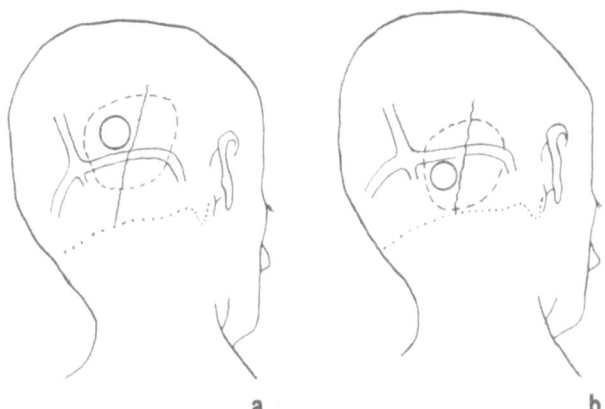

Fig. 7. The sketches showing the site of opening for endoscopic removal of occipitopolar epidural haematomas depending on occipital (a) or posterior fossa (b) spread

of the co-axial draining system, endoscopic dissector, forceps in various modifications[4]. As to liquid chronic haematomas (Fig. 8) endoscopy ensures the visualisation detection, of the dissector perforation and incision of membranes (Fig. 8c–e) controlling blood accumulations for effective drainage of united cavities. Moreover, endoscopic techniques also ensure the submembraneous approach for separation of visceral haematoma membrane from the arachnoid destroying the adhesions between them.

Subdural hygromas. Endoscopic technique allows one to remove the fluid confined by cortical gyri beyond the cranial opening and ensures the drainage of the cavity into subarachnoid space (usually lateral cistern) through an arachnoid opening.

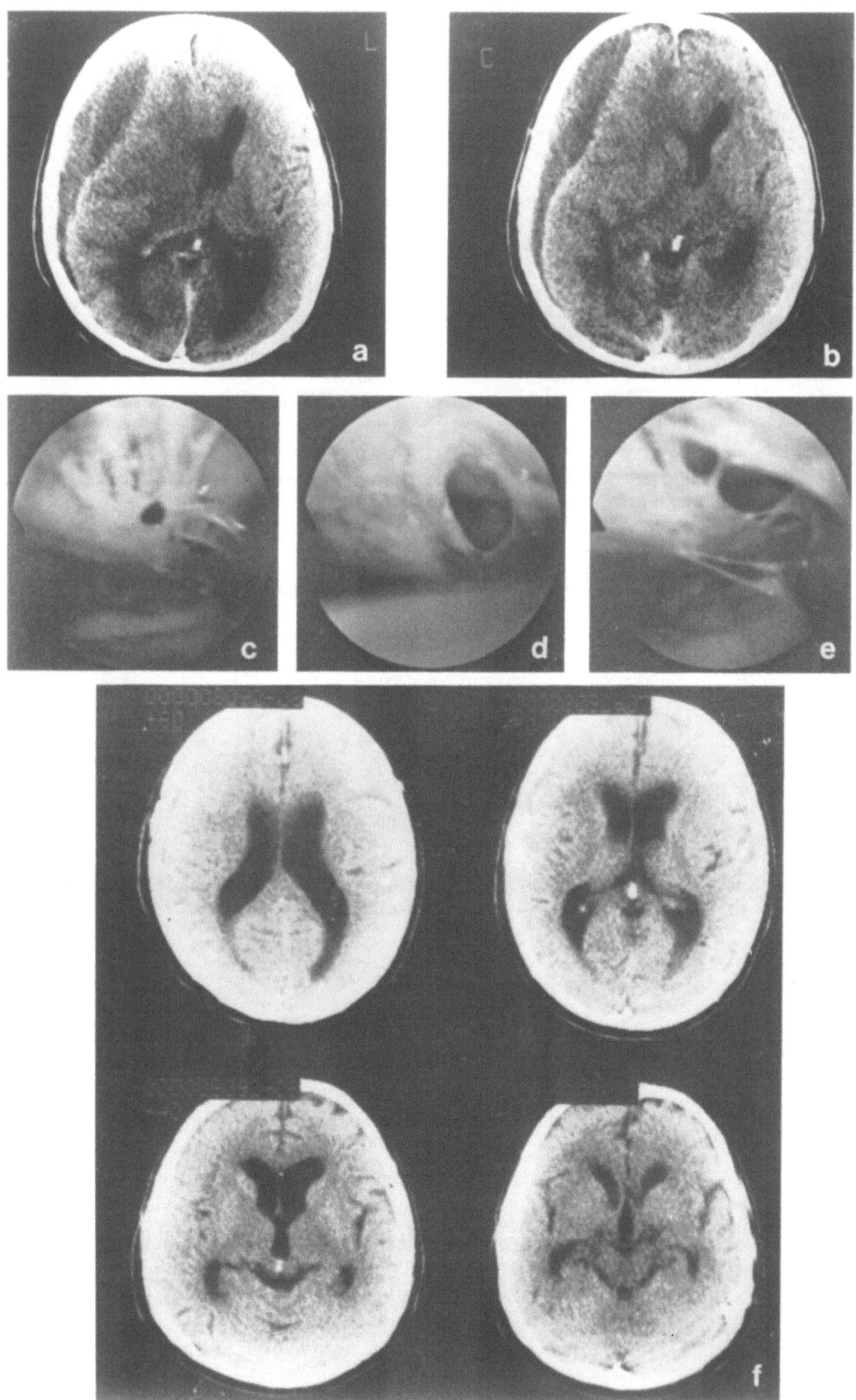

Fig. 8. The significance of the endofiberoscopic removal of liquid chronic subdural haematomas. (a, b) Preoperative CT pictures showing septated form of haematoma; (c, d, e) endoscopic perforation and incision of intercavital membrane by endoscopic dissector; (f) serial postoperative, CT control

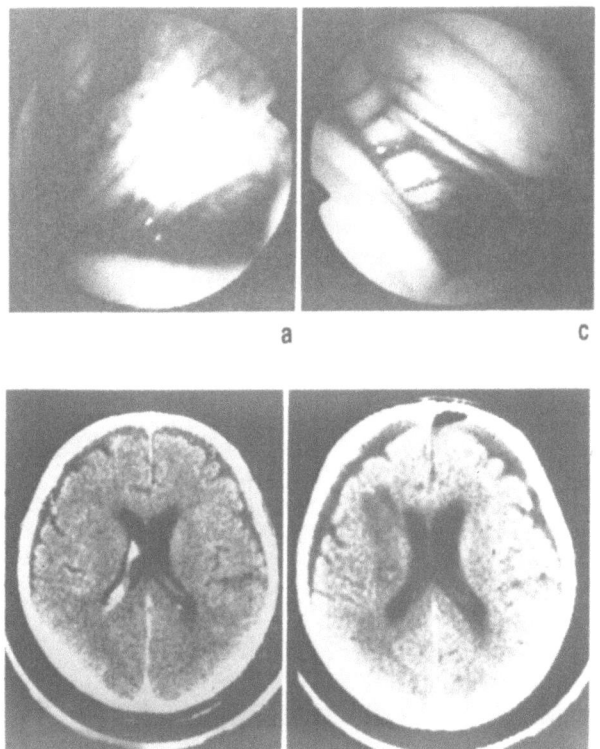

Fig. 9. Endoscopic and CT pictures before (a, c) and after removal (b, d) of clotted haematoma from the lateral ventricle. After clearance of ventricle from blood the atrial veins are detected (c)

Endofiberscopy provides also for the evacuation of *intracerebral haematomas* in toto.

Intraventricular haemorrhages (Fig. 9). Knowledge of endoscopic ventricular topography allows one to remove the clotted and mixed haematomas from the lateral and third ventricles, especially around the interventricular foramen of Monro. This is a typical localization of blood in intraventricular haemorrhages[1,4].

Multiple haematomas. The conservative character of the operative technique ensures the removal of large bilateral haematomas shift.

General results are summarized in Table 2. Relapsed haematomas were removed endoscopically. Death rate (18 % at whole) chiefly depends on the severity of the patient's clinical state, less on the type of haematomas and does not depend on their configuration (Fig. 10). Nevertheless multiple haematomas increase the mortality rate.

Failures. In several cases of acute subdural haematomas it was difficult to reach endoscopically the layer between big haematomas and the brain which penetration resulted in difficulties of clot removal. This required an enlargement of the cranial opening in 1 case. Osteoplastic trepanation has been performed also in 9 cases with prolapsed brain or large cerebral lacerations.

Complications such as damage to brain tissue or its vessels by the endoscope and wound suppuration were not recorded.

Discussion

The experience with more than 200 operations allows one to formulate **indications** for endoscopic haematoma neurosurgery. These are the indications for surgery in general with the exception of contraindications.

The *contra-indications* are the following (corresponding to contemporary technical possibilities):

1. Widespread brain laceration
2. Large bleeding vessel
3. Brain prolapse
4. Calcification of a haematoma

The last point has only academic significance.

Table 2. *Endoscopic Surgery of Traumatic Intracranial Haematomas (General Results)*

Haematomas	Operated on	Volume (ml)	Relapse	Died
Epidural	21	50–250	1[a]	—
Subdural	122	60–200	4[a]	20
Hydromas	6	60–120	—	4
Intracerebral	8	40–80	—	—
Intraventricular	3	15–30	—	2
Multiple[a]	20	up to 230	—	7
	180 (201 operations[a])		5 (2, 5 %)	33 (18, 3 %)

[a] Bilateral 16; relapse surgery 5.

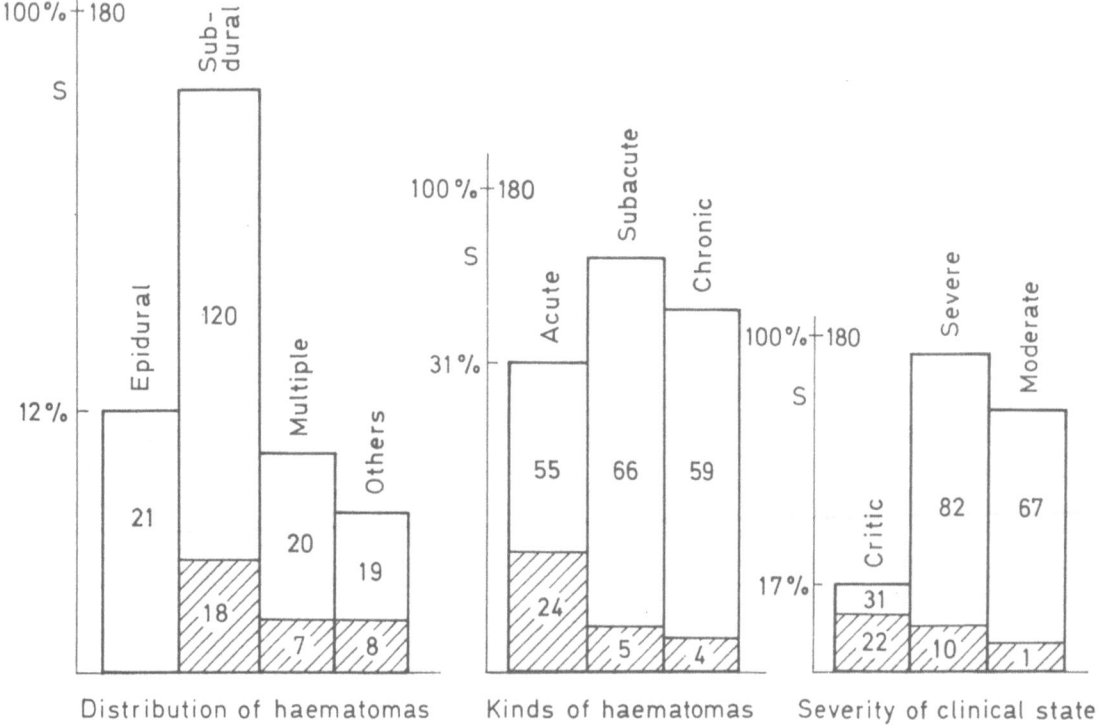

Fig. 10. Diagrams, delineating the mortality rate (hatched parts of columnes) after endoscopic removal of various forms of traumatic intracranial haematomas

The *main advantages* of endoscopic surgery of intracranial haematomas are:

1. Reduction of the surgical trauma by
   — restriction of dural opening;
   — minimal retraction of brain structures during surgery;
   — small skin incision.
2. Shortening of the period of inpatient care and cost reduction.
3. Possibility to operate without assistance as well as outside neurosurgical clinics (disasters, road accidents), combining operative endoscopic with diagnosis and surgery.

Low recurrence (2.5%) is possibly the result of the application in most cases of the delayed non-surgical cranioplasty technique described, providing for the insertion of a wide-bore tube for intensive short-time drainage with ultrasonic and endoscopic control of its effectiveness. The exact definitive bone disc replacement provides an anatomical completion of the operation, skull closure. This technique also prevents the intracranial bone flap migration and postoperative extradural haematomas. Dealing with the relapse rate

of intracranial and especially subdural haematomas, the comparative assessment of our results and summary statistics of "blind" burr-hole evacuation of only liquid subdural haematomas (relapse rate was 6–24%[1,5]) shows the significant decrease of relapse rate using endoscopic techniques. This technique ensures double visual control: 1) during surgery for precise manipulations; and 2) postoperative control of the effectiveness of drainage and its duration.

And finally, the special advantages of endoscopic evacuation of chronic subdural haematomas: 1) incision of membranes restricting the evacuation of blood accumulation for effective drainage of several cavities; 2) withdrawal of remaining clots; 3) separation of thick inner membrane from cortical arachnoid with removal of connective tissue adhesions between them.

Thus, endofiberoscopic techniques optimize the operative surgery of intracranial haematomas.

References

1. Cordobes F, Fuente M, Lobato RD *et al* (1983) Intraventricular haemorrhage in severe head injury. J Neurosurg 58: 217–222
2. Harders A, Eggert HR, Weigel K (1982) Behandlung des chronischen Subduralhämatoms mit externer geschlossener Drain-

age. Bericht über 100 konsekutive Fälle. Neurochirurgia 25: 147–152
3. Karakhan VB (1988). Experience with using intracranial endoscopy in neurotraumatology. Vestn Khir 140(3): 102–108 (English abstract)
4. Karakhan VB (1992). Endofiberscopic intracranial stereotopography and endofiberscopic neurosurgery. In: Bauer BL, Hellwig D (eds) Minimally invasive neurosurgery — MIN. Acta Neurochir (Wien) Suppl 54: 11–25

5. Markwalder TM, Reulen H-J (1986) Influence of neomembranous organization cortical expansion and subdural pressure on the post-operative course of chronic subdural hematoma—an analysis of 201 cases. Acta Neurochir (Wien) 79: 100–106

Correspondence: V.B. Karakhan, M.D., Department of Neurology and Neurosurgery, Moscow Medical Stomatological Institute, Delegatskaya str. 20/1, 103473 Moscow, Russia.

Acta Neurochir (1994) [Suppl] 61: 92–97

Endoscopic Stereotactic Interventions in the Treatment of Brain Lesions

L. Zamorano, C. Chavantes, and **F. Moure**

Department of Neurosurgery, Wayne State University, Detroit, MI, USA

Summary

Image-guided stereotaxis is an accurate and safe method of directing therapy to target volumes defined in two-dimensional (2D) multiplanes or three-dimensional (3D) perspectives using computer reconstruction of image data. The major limitations of stereotactic techniques are a lack of intraoperative visualization and direct monitoring of the procedures, and changes of intracranial coordinates after decompression of cystic lesions or aspiration of cerebrospinal fluid in the management of intraventricular lesions. Endoscopic laser stereotaxis (ELS) involves integration of rigid-flexible endoscopy and Nd-YAG laser to 3D-2D multiplanar image-guided stereotactic procedures. The major advantages of ELS include: minimally invasive (burrhole or small craniotomy surgery), direct intraoperative visualization, hemostasis, evacuation or resection assessment, and wide exploration of intracranial cavities or ventricles. ELS has been used in the management of 152 clinical cases including biopsy, aspiration, resection and internal decompression of deep and subcortical intracranial lesions, and different types of fenestration procedures. Image-guidance combined with endoscopic techniques may offer a safe, accurate alternative to conventional neurosurgical procedures in treating small solid, cystic, and intraventricular lesions as well as fenestration procedures.

Keywords: Endoscopy; fiberoptic; stereotaxis; laser; imaging.

Introduction

Image-guided stereotaxis is an accurate and safe way to approach intracranial lesions, allowing treatment to be directed at target volumes defined in two-dimensional (2D) and three-dimensional (3D) computer reconstructions of image data[3,6,7,15]. With 2D multiplanar or 3D image processing, the size, shape, and main axis of intracranial lesions can be determined along with the important surrounding anatomical structures[5,8,10,14,17]. Image preplanning goals include selection of the safest and optimal approach for diagnosis and treatment of intracranial lesions[9,11,13,14]. This information can be transposed into stereotactic space with mathematical accuracy. Adequate methodology with 1 mm accuracy can be achieved on both the x and y axis and 1.5 mm in z

axis[6]. Nevertheless, major criticisms and limitations of image-guided stereotaxis include the lack of direct visualization and intraoperative monitoring of the procedures, morbidity and mortality due to hemorrhages associated with intraoperative blindness, and changes in intracranial coordinates after aspiration of cystic cavities or intraventricular lesions. To overcome some of these limitations in addition to increasing the scope of clinical usefulness of image-guidance to a wider range of intracranial pathology, we have adapted flexible and rigid endoscopy and the Nd-Yag laser to stereotaxis for management of cystic and intraventricular lesions and fenestration procedures[9,11,12,16].

Material and Method

Instrumentation

The basic instruments used are described in Table 1. These include endoscopic instrumentation (rigid-flexible), a stereotactic frame (Zamorano-Dujovny Multipurpose Neurosurgical Localizing Unit, Fischer, Freiburg, Germany; hereafter, Z-D) and a cannula adapter for a stereotactic arc. A rigid neuroendoscope was developed for diagnostic and therapeutic endoscopy under stereotactic conditions (Karl Storz, Tuttlingen, Germany). The instrument consists of a rigid cannula with an outer diameter of 8 mm and a scale for depth measurements. This cannula can be inserted using the adapter for the stereotactic arc to achieve the image-defined trajectory; an atraumatic mandrin is used to prevent damage to the surrounding brain. Once the desired depth is reached level the mandrin is replaced by a one-piece instrument that includes the optical system (0 degree optics and illumination), separate suction and irrigation channels, and a 2 mm working channel for rigid instrumentation. With a stopcock the irrigation channel can be used simultaneously to harbour a laser fiber or flexible instrumentation.

Optics of 30 and 90 degrees may be useful in some cases. Endoscopic rigid instrumentation includes different biopsy forceps (punch cup, alligator, etc), hook scissors, and bipolar coagulating electrodes. Fiberoptic neuroendoscopy was performed with a flexible Olympus fiberendoscope (ENT-1T10) (Olympus Corporation, Strongsville, OH) and flexible instrumentation similar to the rigid endoscope. A CCD videocamera attached to the neuroendoscope's

Table 1. *Endoscopic-ND-YAG Laser Stereotaxis (ELS)*

Requirements

Image data	CT, MRI, DA
Stereotactic system	Z-D (Zamorano-Dujovny) Multipurpose Neurosurgical Localizing Unit (Fischer, Freiburg, Germany)
Rigid endoscope	Chavantes-Zamorano Endoscope (Karl Storz, Tuttlingen, Germany)
Rigid instrumentation	grasping forceps, scissors, suction needle, bipolar probe
Flexible endoscope	Olympus ENT-1T10
Flexible instruments	grasping forceps, bipolar probe, suction needle
Xenon light source (300 w)	Karl Storz, Tuttlingen, Germany
Videocamera CCD	Karl Storz, Tuttlingen, Germany
Nd-YAG laser	Sharplan 2100, Allendale, New Jersey
Contact laser fibers	400, 600, 800 micron sculpted laser fiber; conical, hemispherical, dual effect (Sharplan Lasers, Allendale, New Jersey)

optic allows display on a television monitor. A 600 micron Nd-Yag laser disposable contact fiber (Sharplan Lasers, Inc., Allendale, New Jersey) can be introduced through the irrigation channel of both rigid and flexible endoscopes for coagulation, cutting, and vaporization of lesions.

Technique

Endoscopic laser stereotaxis is a three-step process: 1) image data acquisition, 2) 3D-2D image processing for stereotactic planning, and 3) intraoperative stereotactic endoscopic procedure.

Image Data Acquisition

A computed tomography (CT)—and magnetic resonance imaging (MRI)—compatible ring (Z-D base ring, F.L. Fischer, Freiburg, West Germany) is fixed to the patient's head in the standard low position for supratentorial, diencephalic, and mesencephalic lesions. An inverted high position is used for lesions located in the infratentorial region or near the skull base. The ring is fixed to the patient's skull with three or four pins that can be placed on any position to allow an unobstructed intraoperative surgical approach. The ring defines the stereotactic coordinate system; the origin is the center of the ring and coordinates are defined as x (right-left), y (anterior-posterior), and z (superior-inferior) to this origin[13]. Image data acquisition is performed with the fixed ring. To interface the image data with the reference system, different approaches can be used depending upon the image modality, i.e. tomographic vs projective. In the case of tomographic imaging (CT or MRI) there are two main approaches: one alternative is the use of localizers attached to the ring which provide landmarks for stereotactic localization. The other approach is the use of an adapter to the CT table or MRI coiler to bring the ring isocentric to the CT gantry or MRI coiler. In projective images (X-ray, digital angiography, DA), thin fidutial radiopaque markers contained in four transparent plates are fixed to the ring; a special algorithm allows coordinates measurements with the center of the ring as absolute origin, independent of source-film distance or angulation[15].

3D-2D Image Processing for Stereotactic Planning

Basic 2D surgical preplanning can be performed at the CT or MRI console with available software. Two-dimensional multiplanar preplanning can be performed with axial and reformatted views in coronal, sagittal, paraaxial, or oblique planes on CT images and with axial, coronal, and sagittal planes using MRI. The location, size, shape, main axis, and anatomical and vascular relationships of the lesion can be visualized. Ideally, stereotactic procedures involve multidimensional 3D and 2D multiplanar interactive surgical preplanning. A "real-time" image-based surgical planning utilizing multimodality imaging (CT, MRI, DA) was developed at our institution using a Sun 4/370 workstation and specially developed hardware and software that provide 2D multiplanar and 3D views of the lesion for stereotactic procedures (NSPS, Stereotactic Neurosurgical Planning System, Wayne State University). The 2D multiplanar view includes reformatted vectors at any arbitrary plane of the patient's lesion such as sagittal, coronal, transverse, paraaxial, and free-tilt views, including oblique vectors. The 3D software includes features for extracting a view of the target volume localized by automatic segmentation, thresholding, manual contouring, and boundary detection. Both 2D and 3D menus include "location" functions with "real-time" coordinate measurements in stereotactic conditions and surgical trajectory definition capabilities as well as statistical functions for measuring distances, angles, areas, and volumes[10,14].

A combined interactive 3D-2D menu allows simultaneous display of selected a trajectory, final optimization, and multiformat 2D display of free-tilt reformatted images perpendicular to selected trajectory of the entire target volume corresponding to the surgeon's eye-view perspective. For ELS preplanning, an automatic thresholding or manual contouring technique is used to delineate treatment areas on each of the axial slices. Treatment volume is calculated and displayed in any desired perspective. Two dimensional multiplanar reformatted images are generated on coronal, sagittal, paraaxial, or oblique planes. The surgical trajectory of the rigid cannula of the endoscope is selected in 3D and 2D images, taking into consideration the tumor's axis, shape, size, number of loculations, vascularity, and surrounding anatomy. Generally, the surgical trajectory will follow the tumor's main axis unless anatomical or vascular considerations contraindicate such an approach. With the interactive 3D and 2D menu, optimization and simulation of the procedure is performed. Finally, a batch of reformatted images is generated at any desired spacing interval that corresponds to the planes perpendicular to the endoscope's trajectory. The aim is to define the lesion and its anatomical relationships at different depths of introduction of the endoscope, and to select the best trajectory for the rigid cannula.

In summary, two-point coordinates that define the optimized trajectory are generated at the time of preplanning and are transposed into stereotactic space by using an algorithm that calculates angles of the aiming device. The multiple slices corresponding to the endoscope's perspective are generated and displayed into a video or computer monitor located in the operating room.

Intraoperative Stereotactic Endoscopic Procedure

Aiming Device Settings

In the case of the Z-D aiming device the semi-arc device can mounted on any four quadrants of the base ring (anterior, right, posterior, or left) according to the surgeon's preference, giving an unobstructed surgical approach. The system is arc-centered and the target x, y, z coordinates can be set directly on the device. With a special algorithm a trajectory is defined by calculating two angles at specific mountings. The probe holder on the aiming device adjusts the instrument depth above or below the target. An arc adapter is used to hold the endoscopic cannula in place.

Endoscopic Procedure

By using a Mayfield adapter the patient is positioned as necessary (supine, semilateral, lateral, prone, semiprone, etc), rotation as well as flexion-deflexion. After the patient's head is shaved at the entry level, the head and ring are completely covered with sterile drapes. At the level of the ring where the localizing unit will be mounted, it is advisable to use transparent sterile drape. From now on, the procedure will be performed under completely sterile conditions which is fundamental in the management of intracranial lesions.

The sterile localizing device is mounted. Under local anesthesia, a burrhole or a 2 cm diameter craniotomy is performed on the image-defined entry area (transposed by the two calculated angles of the device). The dura is incised in a cruciate fashion. The 2 mm stereotactic probe is inserted up to the desired depth (usually the center of a cystic lesion) to obtain diagnostic tissue and fluid for cytologic or microbiologic studies. After removing the probe, the metalic cannula (8 mm) with an obturador is inserted following the same trajectory through the adapter connected to the stereotactic arc. Rigid endoscopic optics is then introduced, and continuous suction and irrigation are connected. Irrigation is performed with 37 centigrade saline solution controlled by a pump activated by a switchpedal. The fluid component of lesions such as hematomas, abscesses, cystic tumors, etc., are then easily evacuated under direct visualization. Continuous irrigation keeps the optical system clean. Through the working channel, biopsy forceps can be used to obtain diagnostic tissue, scissors can be used to incise septations, and the Nd-YAG laser fiber can be used for coagulation, incision, or vaporization.

In some cases of a large cystic or intraventricular process, the use of the flexible endoscope provides for tissue inspection, resection, or vaporization in otherwise inaccessible lesions. This is accomplished by tip flexion-deflexion and rotation of the endoscope. With flexible instrumentation, complete control of tumor cavities can be achieved. The Nd-YAG laser fiber can be used for further hemostasis or resection.

Clinical Experience

Our initial clinical experience with ELS on 152 patients has been encouraging. The technique has proved to be accurate, safe, and resulted in low morbidity (see Figs. 1–4). Our patients had mainly cystic or small solid intraaxial lesions (tumors, hematomas, benign cysts, abscesses, etc.); some had intraventricular lesions. Histological diagnosis has been obtained in all cases as well as immediate internal decompression (Table 2). Operative mortality was seen in one patient. This patient had an uneventful decompression of a large cystic metastatic adenocarcinoma of the lung, but respiratory arrest occurred secondary to probable intraoperative seizure or hypoxia. He was

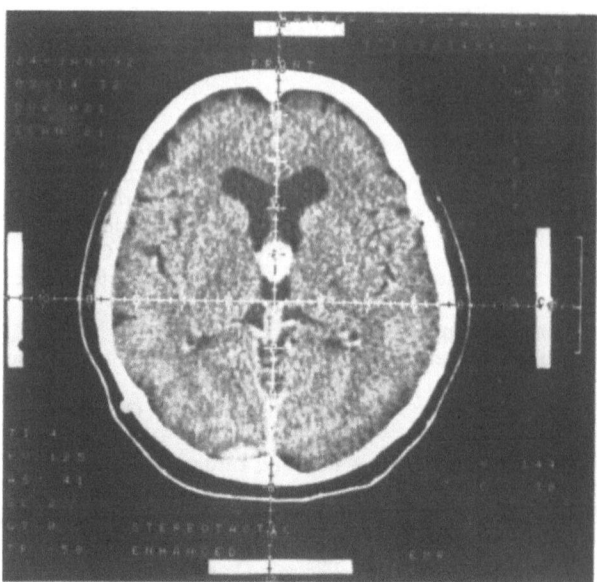

Fig. 1. Preoperative CT scan showing hyperdense colloid cyst of the anterior third ventricle causing hydrocephalus

intubated and ventilated but died three days postoperatively due to irreversible brain hypoxic damage.

Discussion

Image-guided stereotaxis has proved to be an accurate and safe method to approach intracranial lesions,

Table 2. *Endoscopic ND-YAG Laser Stereotaxis (ELS): Indications and Diagnosis on 152 Patients*

Biopsies and/or resection/decompression of lesions	
Tumors	99
Glioblastoma multiforme	25
Anaplastic astrocytoma	18
Astrocytoma	9
Mixed oligo-astrocytoma	6
Craniopharyngioma	11
Metastasis	21
Plexus papilloma	2
Pinealoma	4
Ependymoma	3
Colloid cysts	6
Radionecrosis	6
Hematomas	12
Abscesses	5
Criptic avm	3
Air	1
Fenestrations	
Arachnoidal cysts	8
Third ventriculostomy	4
Septostomy	3
Ventricular cysts	5

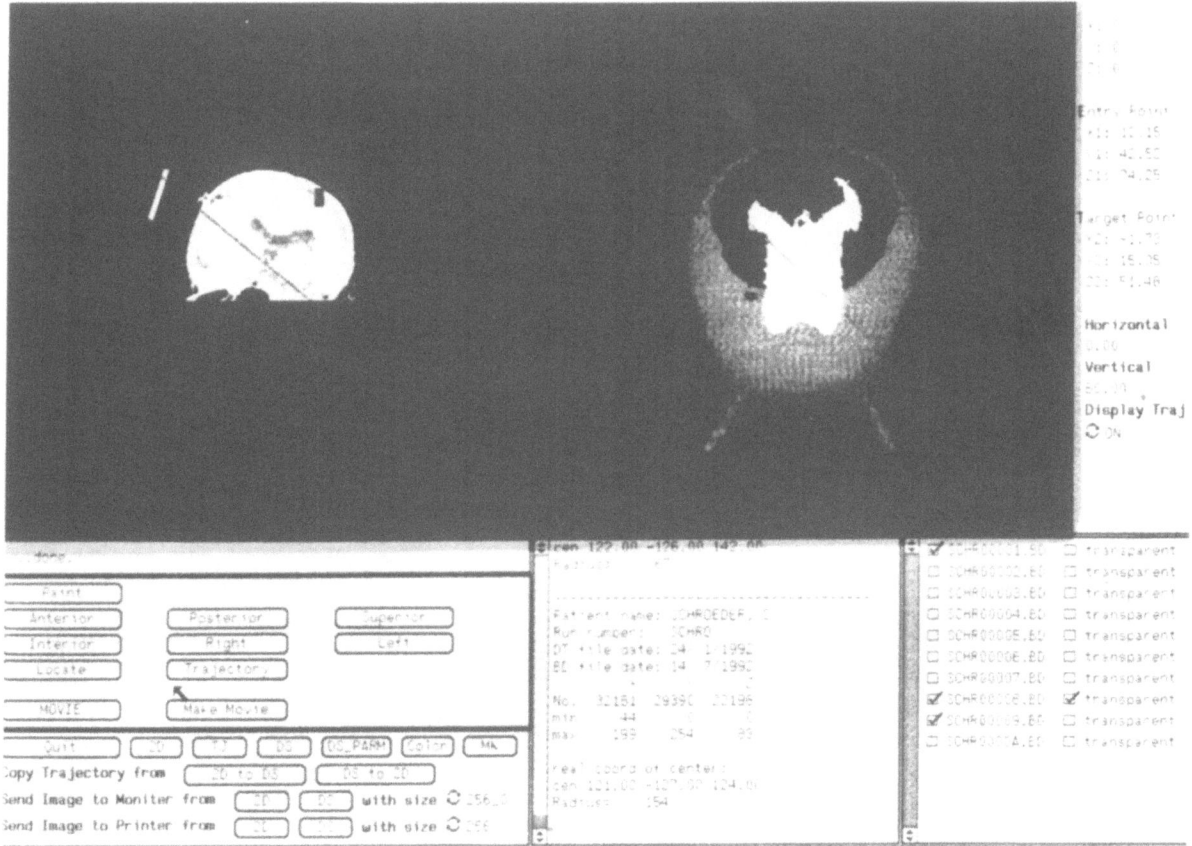

Fig. 2. Image simulation of endoscopic surgical approach; left: oblique 2D view showing planned trajectory of rigid endoscope through lateral ventricle, foramen of Monro, and colloid cyst. Right: same trajectory in 3D image showing skull, ventricle, and cyst

directing therapy to target volumes defined in 2D and 3D using computer reconstruction of image data. With 2D-multiplanar or 3D image reconstructions, size, shape, and the main axis of intracranial lesions can be determined to select the best approach for diagnostic and therapeutic purposes. This volumetric information can be transposed into stereotactic space. Integration of endoscopy with stereotaxis allows direct intraoperative visualization and monitoring of stereotactic procedures[9, 11, 12, 16]. Main indications of ELS technique include biopsy, internal decompression of cystic tumors or resection of small tumors (i.e., malignant gliomas, metastases, craniopharyngiomas), benign cystic lesions such as arachnoidal cysts, colloid cysts, abscesses, and intracerebral hematomas.

Other important indications for ELS are intraventricular processes, such as intraventricular hemorrhage, tumors, and fenestrations. Flexible endoscopes can also be used as a sole instrument in the management of epi- or subdural collections, such as chronic subdural hematomas, abscesses, etc., and in the management of intraventricular pathology such as hemorrhage and hydrocephalus. Hemostasis can be achieved by laser photocoagulation. The potential application of this technology for neurosurgery is significant. Technically some of the features of the Z-D stereotactic system make it appear as a very useful tool when coupled with endoscopy in biopsy, resection, internal decompression, and fenestration procedures[13]. Multimodality imaging compatibility, freedom to choose intraoperative patient positioning, unobstructed surgical approach by multiple alternative mounting of the aiming device, complete intraoperative sterility, and accurate target and trajectory transposition are some of the features. Also, we believe the use of image processing capabilities, such as NSPS, represents an enormous advantage in any intracranial procedure, allowing the neurosurgeon to "simulate" different surgical approaches and select the best one based on 2D and 3D images. Additionally, once a plan is selected this can be accurately transposed into the operating room, providing a "pre-

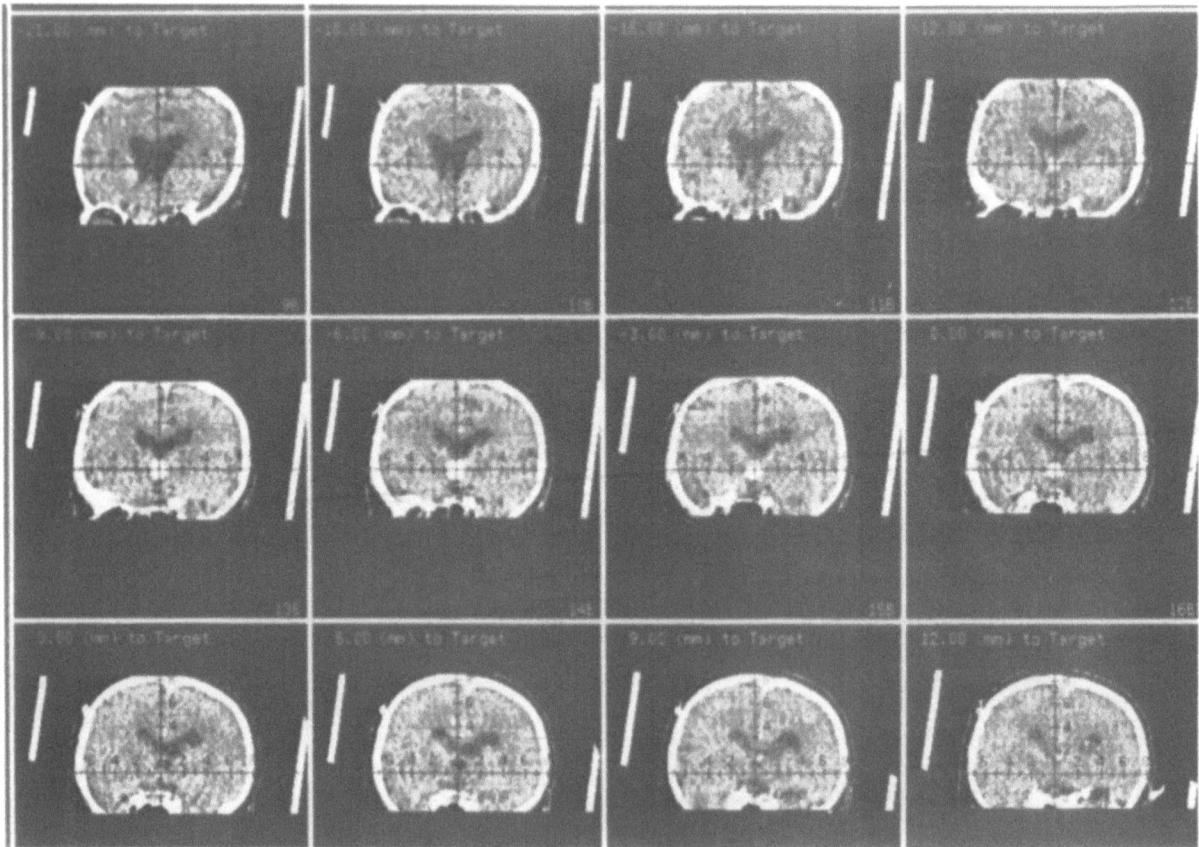

Fig. 3. Series of slices perpendicular to trajectory corresponding to endoscope's eyeview perspective at different depth

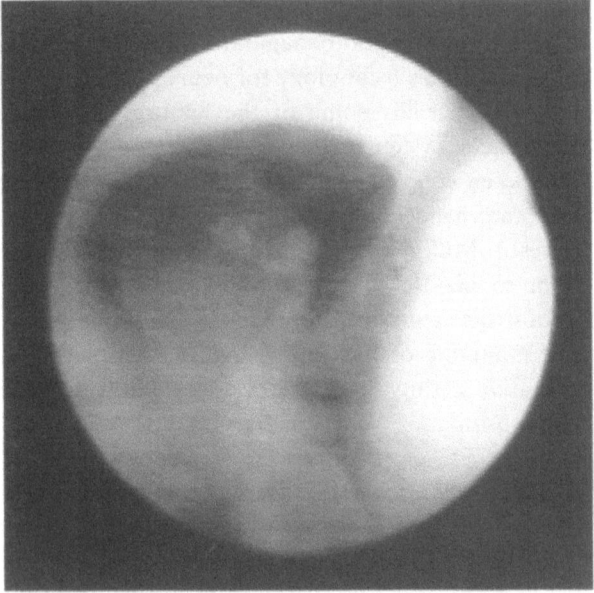

Fig. 4. Intraoperative endoscopic view showing colloid cyst at foramen of Monro, choroid plexus, and anterior septal vein

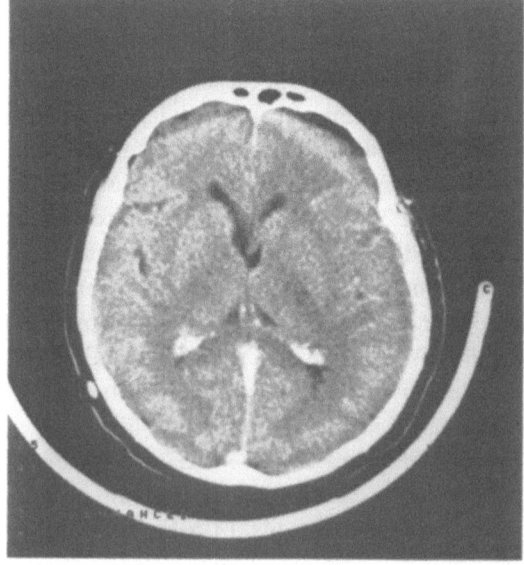

Fig. 5. Postoperative CT scan showing abscence of cyst and resolution of hydrocephalus

known" anatomy at any specific depth of the selected surgical trajectory[10,14]. Ultimately, all this represents an initial step towards the implementation of real-time interactive endoscopic stereotaxis.

We are currently working with an opto-electronic system based on LEDs and three infrared cameras to localize on real-time the position and trajectory of the rigid endoscope. The position and orientation of the endoscope with respect to the precaptured CT/MRI/DA images can be displayed in 2D and 3D presentation on a computer monitor. This electronic coupling can then serve as a precise guide to the performance of an actual endoscopic procedure.

Conclusion

Endoscopic-Nd-YAG Laser Stereotaxis (ELS) is a promising technique for resection (cytoreduction) of some image-defined intracranial tumors and may be valuable in the management of malignant tumors. The impact of this therapeutic model as adjuvant to other volume-depending treatment modalities, such as brachytherapy, hyperthermia, phototherapy, etc., needs to be assessed[2,4,5,8,17]. Integration of image-guided lasers to treat intracranial tumors allows precise removal and vaporization of these lesions[1]. The impact of this technology on patient survival needs to be evaluated. However, beyond the precise "cutting" or "vaporization" effect on tissues, and beyond the photocoagulator effect (both responsible for lasers being considered precise surgical tools) applications of laser technology in endoscopic stereotaxis will result in new diagnostic and therapeutic modalities. The combination of laser beam-tumor interaction and image guidance may provide new ways to treat tumors, such as accurate photodynamic therapy, and photothermally-induced hyperthermia[2,4]. Future advances in miniaturization of fiberoptics for endoscopy, imaging, electronic control, and in lasers technology will undoubtedly enhance and optimize the use of endoscopic laser stereotaxis.

References

1. Beck O, Wilske J, Schonberger G (1979) Tissue changes following application of lasers to the rabbit brain. Results of the CO_2 and Nd YAG lasers. Neurosurg Rev 1: 31–26

2. Bleehan N (1982) Hyperthermia in the treatment of cancer. Br J Cancer 45: 96–100

3. Kelly P, Kall B, Goerss S (1986) Results of computer-assisted stereotactic laser resection of deep-seated intracranial lesions. Mayo Clin Proc 61: 20–27

4. Mang T, Dougherty T (1986) Use of the Nd-YAG laser for hyperthermia induction as an adjunct to photodynamic therapy. Laser Surg Med 6: 237

5. Zamorano L, Dujovny M, Malik G, Yakar D, Mehta B (1987) Multiplanar CT-guided stereotaxis and I-125 interstitial radiotherapy: image-guided tumor volume assessment, planning, dosimetric calculations, stereotactic biopsy and implantation of removable catheters. Appl Neurophysiol 50: 281–286.

6. Zamorano L, Dujovny M, Malik G, Mehta B, Yakar D (1987) Factors affecting measurements in computed-tomography guided stereotactic Procedures. Appl Neurophysiol 50: 53–56

7. Zamorano L, Martinez-Coll, Dujovny M (1989) Transposition of image-defined trajectories into arc-quadrant centered stereotactic systems. Acta Neurochir (Wien) [Suppl] 46: 95–103

8. Zamorano L, Dujovny M, Yakar D, Malik G, Chavantes C, Mehta B (1989) Multiplanar image-guided stereotactic brachytherapy with iodine 125. In: Dyke P et al (eds) Neurosurgery: state of the art reviews, Vol 4, Suppl 95–103. Hanley and Belfus Philadelphia, pp 95–103

9. Zamorano L, Chavantes C, Dujovny M, Malik G (1989) Image-guided stereotactic resection of intracranial lesions: endoscopic and laser technique. In: Dyke P et al (eds) Neurosurgery: state of the art reviews, Vol 4, Suppl 105–118. Hanley and Belfus, Philadelphia, pp 105–118

10. Zamorano L, Dujovny M, Ausman J (1989) Three-dimensional/two-dimensional multiplanar stereotactic planning system: hardware and software configuration. Applications of digital Image Processing XII. SPIE, Vol 1153, pp 552–567

11. Zamorano L, Chavantes C, Dujovny M, Malik G, Ausman J (1990) Image-guided endoscopic laser stereotaxis (ELS). Stereotact Funct Neurosurgery 54 and 55: 421

12. Zamorano L, Chavantes C, Dujovny M, Ausman J (1990) Endoscopic laser stereotaxis: indication for cystic and intraventricular lesion. SPIE, Vol 1200. Laser surgery, pp 253–271

13. Zamorano L, Dujovny M (1991) ZD Multipurpose neurosurgical image-guided localizing unit: experience in 103 consecutive cases of open stereotaxis. SPIE, Vol 1428, pp 30–51

14. Zamorano L, Dujovny M (1991) Multidimensional tomographic image processing for surgical planning of meningiomas. In: Schmidek HH (eds) Meningiomas and their surgical management. Saunders, Philadelphia, pp 163–170

15. Zamorano L, Bauer-Kirpes B, Dujovny M, Malik G, Ausman J (1991) Application of multimodality imaging stereotactic localization in the surgical management of vascular lesions. Acta Neurochir (Wien) [Suppl] 52: 67–68

16. Zamorano L, Chavantes C, Dujovny M, Malik G, Ausman J (1992) Stereotactic endoscopic interventions in cystic and intraventricular brain lesions. Acta Neurochir (Wien) [Suppl] 54: 69–76

17. Zamorano L, Bauer-Kirpes B, Dujovny M, Yakar D (1992) Dose planning for interstitial irradiation. In: Kelly P (ed) Computers in stereotactic surgery. Blackwell, Oxford, pp 280–291

Correspondence: Lucia Zamorano, M.D., Ph.D., Department of Neurological Surgery, Wayne State University, School of Medicine, 4201 St. Antoine, 6E, Detroit, Michigan 48201, U.S.A.

Acta Neurochir (1994) [Suppl] 61: 98–101

Stereotactic Endoscopic Resection of Angiographically Occult Vascular Malformations

T. Otsuki, H. Jokura, N. Nakasato, and **T. Yoshimoto**

Department of Neurosurgery, Miyagi National Hospital and Tohoku University School of Medicine, Sendai, Japan

Summary

Total resection of angiographically occult vascular malformations was performed in four patients using stereotactic open-system endoscopy. The lesions were relatively small but localized deep in the brain or involving eloquent cortex. Fiber-guided YAG laser coagulation was efficient for haemostasis and no significant complication was seen in any case. Endoscopic stereotactic laser surgery by open-system endoscopy is considered a safe and promising treatment for small AOVM's located in deep or eloquent neural structures.

Keywords: Occult vascular malformation; endoscopic surgery; stereotactic technique.

Introduction

Advent of CT and MRI discloses an increasing number of angiographically occult vascular malformations (AOVM's), which may cause haemorrhages, seizures and progressive neurological deficits[4,7,8,11,12,15,17,22,23]. Although the natural prognosis of AOVM's is not fully understood[11,16,24], surgical excision is generally accepted for definite diagnosis and prevention of recurrent haemorrhage[5,13,16,19,20].

Surgical management of AOVM's, however, is not always simple since these lesions tend to be small and located deep in the brain so that the localization of the lesions by conventional surgery may be difficult. Stereotactic radiosurgery, as an alternative to resective surgery, has met with limited success for AOVM's so far[6,13,22]. Stereotactic needle biopsy of AOVM's for diagnostic evaluation may cause massive and fatal haemorrhage on some occasions.

In order to resect small AOVM's with definite accuracy and minimal invasiveness, the authors utilized a new stereotactic endoscopic approach using a stereotactic guiding tube and fine endoscopes[9,10]. This approach allows the surgeon to use various surgical instruments including high power lasers so that endoscopic resection of solid intra-axial lesions becomes possible.

In this report, our initial experience of endoscopic resection in four patients with AOVM's are described.

Operative Technique

The details of the devices and techniques used for this operative procedure were reported previously[9,10]. Briefly, patients underwent CT scanning of 2 mm slices with their heads fixed in a Leksell stereotaxic frame. XYZ coordination of the center of the lesions is obtained and the trajectory is determined based on reconstructed CT images. The depth and size of the lesions and the angle of the trajectory are measured from reconstructed multiplannar CT images.

A guiding tube, 8 mm in diameter, with an inner stylet is adapted to the distal needle holder of Leksell's stereotaxic apparatus and inserted stereotaxically to the targets through an ordinary burr hole. At the target point, the stylet is removed and either a rigid or flexible endoscope is inserted into the guiding tube (Fig. 1). The eye piece of the rigid endoscope is fixed to the proximal needle holder of Leksell's frame, whereas that of the flexible endoscope to a specially designed holding apparatus is attached to the operating table.

Various flexible microsurgical instruments, such as suction tubes, tumour forceps, dissectors or scissors, are inserted through the side window of the guiding tube and manipulated so that the lesions could be removed after dissection from the surrounding brain tissue. A Nd:YAG laser beam at a power of 10–20 W is transmitted through a quartz flexible fiber, which is inserted into a metal tube and applied for coagulation and haemostasis.

Summary of Cases

Clinical Data

During the past year, four patients with angiographically occult vascular malformations were treated by

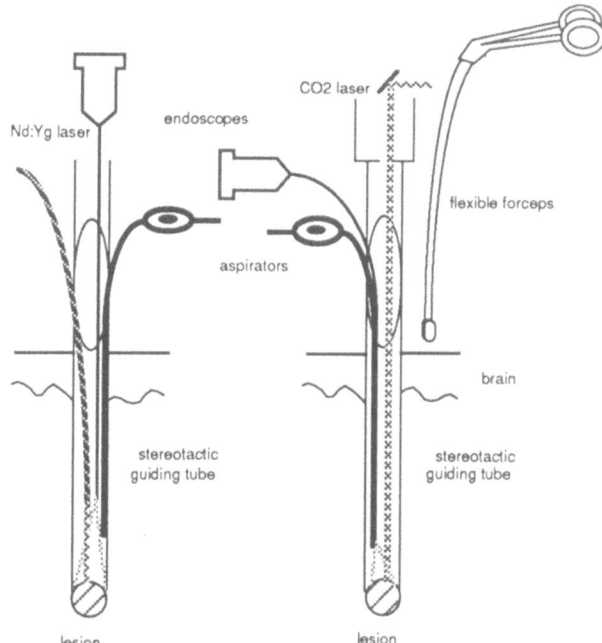

Fig. 1. Basic concept of the stereotactic open-system endoscopy. The stereotactically inserted guiding tube substitutes for the brain retractors to expose the deep-seated lesions, which are visualized by either rigid (left) or flexible (right) endoscope. Various flexible surgical instruments including lasers can be inserted through the open windows of the guiding tube

this method (Table 1). The patients included three women and one man, aged 43 to 60 years old. Clinical symptoms were acute headache or numbness in the hand accompanied by minor haemorrhage. Another two patients were referred to our clinic for histological diagnosis.

Radiological Findings

All the patients underwent CT, MRI and cerebral angiography. The lesions were 10–18 mm in the maximal diameter on CT images and located in the lateral ventricle (Fig. 2), centrum semiovale, sensory cortex and deep frontal lobe. The angiogram showed no evidence of abnormal vascularity in any cases. CT scan

showed homogeneous high density lesions in three and mixed density in one case. Typical mixed high and low signal intensities on MRI were recognized in one case whereas homogeneous signal intensity was shown in the remaining three cases.

Operative Procedure

Burr holes were placed in the parietal or the frontal bone. The distance from the cortical surface to the center of the lesions was 10–45 mm (average 32.5 mm). All the lesions were well-circumscribed and easy to dissect from the surrounding tissue. The intraventricular lesion was detached from the choroid plexus using micro-scissors with YAG laser coagulation after placing cotton pledgets around the lesion to prevent blood accumulation within the lateral ventricle. The operative fields were always kept clear by means of irrigation and suction when required. Total removal was achieved in all cases without experiencing critical intra-operative bleeding.

Pathological Findings

Histological diagnosis was cavernous angioma in three cases. Case 2 was diagnosed as destructed cryptic AVM with a haematoma.

Complications

No significant complications were experienced. Postoperative minor headache was observed in case 1, which disappeared within a week.

Follow-up Results

All the patients have maintained their original social activities without occurrence of any new symptoms for the follow-up period (2.8–3.7 y). Follow-up radiological studies have shown no recurrence of haemorrhage nor evidence of residual lesions.

Table 1. *Summary of Cases*

Case	Age	Sex	Symptoms + signs	Location	Size/Depth	Histology
1	60	F	none	lateral ventricle	12/45 mm	cavernous angioma
2	43	F	acute headache	centrum semiovale	18/37 mm	hematoma
3	53	M	acute numbness	sensory cortex	11/10 mm	cavernous angioma
4	55	F	chronic headache	deep frontal lobe	10/35 mm	cavernous angioma

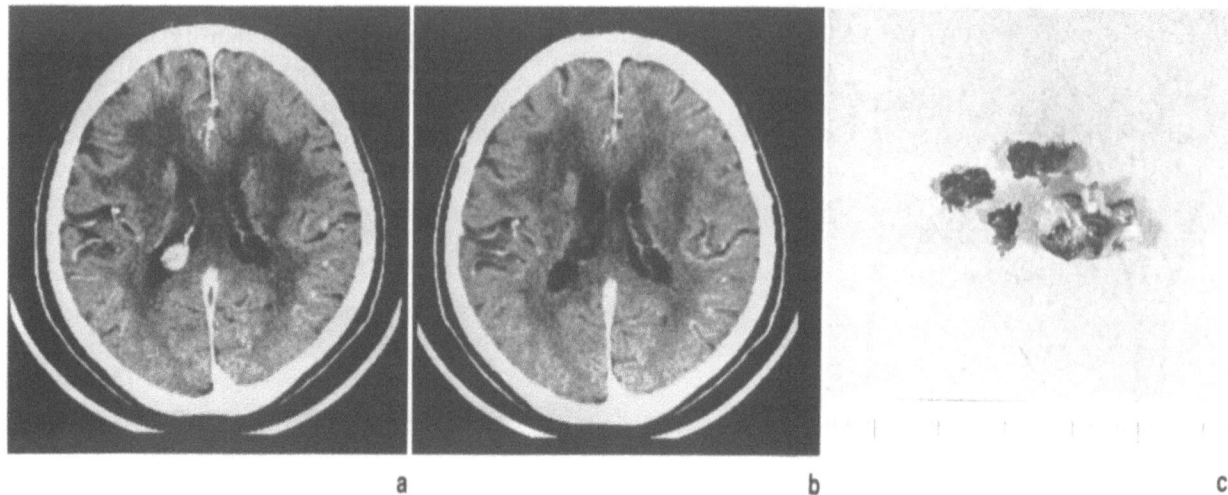

Fig. 2. Pre-operative (a) and post-operative (b) CT scans and surgical specimen (c) of 60 year-old female (case 1). Stereotactic endoscopic resection was performed through a burr hole of the usual size in the parietal bone under general anaesthesia. Fiber-guided YAG laser was used for coagulation and haemostasis. Histological diagnosis was cavernous angioma

Discussion

Angiographically occult vascular malformation (AOVM) is defined as a vascular malformation of the brain which cannot be detected by angiography[7,8]. The pathology of AOVM's may include any type of vascular malformation such as cavernous angioma, venous angioma, teleangiectasia, and true arteriovenous malformation (AVM). Cavernous angiomas are hardly visualized by angiography since they are separated from the main vascular supply. Small or cryptic AVM's with haemorrhage may not be detected by angiography because of compression by adjacent clots or destruction at the time of bleeding.

In this report, the histological diagnosis of the AOVM's resected by endoscopy was cavernous angioma in three cases and nonspecific haematoma in one, the latter of which was considered to be the case of an obliterated cryptic AVM.

Development of CT and MRI discloses increasing numbers of AOVM's incidentally or with minor symptoms such as headache, seizures, or local symptoms caused by haemorrhage. Although AOVM's have been reported to show some characteristic appearances on CT and MR images[12,15] representing haemosiderin deposits and haematomas in different stages of evolution, some neoplasms can also show an identical appearance that is indistinguishable from that of AOVM's[2,18]. The differential diagnosis of these lesions includes intra-axial neoplasms, granulomatous lesions, and even a vascular meningioma[8]. Recent reports suggest that AOVM's do not have a

benign natural history as considered previously[19] and they are prone to cause recurrent haemorrhages and persistent neurological deficits. Complete resection of AOVM's, therefore, is recommended for symptomatic patients if the lesion is accessible[5,13,19,20] although natural prognosis especially the rate of rebleeding of AOVM's is not well known yet[11,16,24].

Although the stereotactic endoscopic approach has been considered an ideal method for resecting small intracranial lesions with minimal invasiveness, endoscopic excision of intraparenchymal lesions has remained difficult[3,14] This is caused by the fact that most conventional neurosurgical endoscopes were modification of urological endoscopes, which were originally developed for surgery in fluid-filled cavities such as the bladder. The endoscopic approach utilized in this report is based on a concept that the stereotactically inserted guiding tube plays the role generally played by retractors in stereotactic microsurgery and that the endoscopes act as microscopes for viewing the deep and confined operating field[7,8]. Since the operating field is kept dry and exposed to air, this endoscopic system can be regarded as an open system versus the closed system previously employed in most neurosurgical endoscopes.

In this report, total resection of AOVM's was performed in four patients using stereotactic opensystem endoscopy. The lesions were relatively small but located in deep or eloquent brain structures. Fiber-guided YAG laser coagulation was efficient for haemostasis and no significant complication was experienced in any cases. Endoscopic stereotactic laser surgery by open-system endoscopy is considered a safe and promising treatment for small AOVM's located in deep or eloquent neural structures.

References

1. Ahmadi J, Miller CA, Segall HD, Parks H, Zee CS, Becker RL (1985) CT patterns in histologically complex hemangiomas. AJNR 6: 389–393

2. Atlas SW, Grossman RI, Gomori JM *et al* (1987) Hemorrhagic intracranial malignant neoplasms; spin-echo MR imaging. Radiology 164: 71–77

3. Auer LM, Holzer P, Asher PW, Heppner F (1988) Endoscopic neurosurgery. Acta Neurochir (Wien) 90: 1–14

4. Chadduck WM, Binet EF, Farrell FW, Araoz CA, Reding DL (1985) Intraventricular cavernous hemangioma: report of three cases and review of the literature. Neurosurgery 16: 189–197

5. Davis DH, Kelly PJ (1990) Stereotactic resection of occult vascular malformations. J Neurosurg 72: 698–702

6. Kondziolka D, Lunsford LD, Coffy RJ, Bissonette DJ, Flickinger JC (1990) Stereotactic radiosurgery of angiographically occult vascular malformations: indications and preliminary experience. Neurosurgery 27: 892–900

7. Lobato RD, Perez C, Rivas JJ, Cordobes F (1988) Clinical, radiological, and pathological spectrum of angiographically occult intracranial vascular malformations, analysis of 21 cases and review of the literature. J Neurosurg 68: 518–531

8. Oglivy CS, Heros RC, Ojemann RG, New P (1988) Angiographically occult vascular malformations. J Neurosurg 69: 350–355

9. Otsuki T, Jokura H, Yoshimoto T (1990) Stereotactic guiding tube for open-system endoscopy: a new approach for the stereotactic endoscopic resection of intra-axial brain tumors. Neurosurg 27: 326–330

10. Otsuki T, Yoshimoto T (1991) A new approach for the endoscopic stereotactic brain surgery using high-power laser. Proceedings of Optical Fibers in Medicine VI; SPIE 1420: 220–224

11. Pozzati E, Giuliani G, Nuzzo G, Poppi M (1989) The growth of cerebral cavernous angiomas. Neurosurgery 25: 92–97

12. Rigamonti D, Drayer BP, Johnson PC, Hadley MN, Zabramski J, Spetzler RF (1987) The MRI appearance of cavernous malformations. J Neurosurg 67: 518–524

13. Sadikot AF, Olivier A, Bertrand G, Podgorsak E, Souhami L (1991) Radiosurgical treatment of angiographically occult arteriovenous malformations. J Neurosurg 74: 346A

14. Seifert V, Gaab MR (1989) Laser-assisted microsurgical extirpation of a brain stem cavernoma: case report. Neurosurg 25: 986–990

15. Shelden CH, Jacques S, Lutes HR (1988) Neurologic endoscopy. In: Schmidek HH, Sweet WH (eds) Operative neurosurgical techniques. Grune and Stratton, Orland, pp 423–430

16. Sigal R, Krief O, Houtteville JP, Halimi P, Doyon D, Pariente D (1990) Occult cerebrovascular malformations: follow-up with MR imaging. Radiology 176-815-819

17. Simard JM, Garcia-Bengochea F, Ballinger WE, Mickle JP, Quisling RG (1986) Cavernous angioma: a review of 126 collected and 12 new clinical cases. Neurosurgery 18: 162–172

18. Sze G, Krol G, Olsen WL, *et al* (1987) Hemorrhagic neoplasms. MR mimics of occult vascular malformations. AJNR 8: 795–802

19. Tagle P, Huete I, Mendez J, DEl Villar S (1986) Intracranial cavernous angioma: presentation and management. J Neurosurg 64: 720–723

20. Tung H, Giannota SL, Chandrasoma PT, Zee C (1990) Recurrent intraparenchymal hemorrhages from angiographycally occult vascular malformations. J Neurosurg 73: 174–180

21. Yoshimoto T, Suzuki J (1986) Radical surgery on cavernous angioma of the brain stem. Surg Neurol 26: 72–78

22. Weil S, Tew JM, Steiner L (1990) Comparison of radiosurgery and microsurgery for treatment of cavernous malformations of the brain stem. J Neurosurg 72: 336A

23. Wharen RE Jr, Scheithauer BW, Laws ER Jr (1982) Thrombosed arteriovenous malformations of the brain. An important entity in the differential diagnosis of intractable focal seizure disorders. J Neurosurg 57: 520–526

24. Wilkins RH (1985) Natural history of intracranial vascular malformations: a review. Neurosurgery 16: 421–430

Correspondence: T. Otsuki, M.D., Department of Neurosurgery, Tohoku University School of Medicine, 1-1 Seinyo-machi, Aoba-ku, Sendai, Japan.

Acta Neurochir (1994) [Suppl] 61: 102–105

Endoscopic Stereotactic Treatment of Brain Abscesses

D. Hellwig, B.L. Bauer, and **W.A. Dauch**

Department of Neurosurgery, Philipps-University Marburg, Federal Republic of Germany

Summary

Treatment of brain abscess is still a subject of controversy. Craniotomy with primary extirpation and resection of the abscess membrane, burrhole craniotomy with puncture or insertion of a drain, marsupialization, or stereotactic aspiration are different therapeutic approaches. As a consequence of our experiences and results with neuro-endoscopic interventions we have introduced endoscopic stereotactic techniques in brain abscess treatment.

Seven patients with brain abscesses were operated on stereotactically using an endoscope. In all cases the abscess contents were aspirated, while the abscess membrane was left in situ. The patients received postoperative antibiotic therapy according to microbial diagnosis. The longest follow-up period was 48 months. Six patients showed a marked improvement of neurological deficit after treatment. One patient died from sepsis caused by a bacterial endocarditis.

The results emphasize that endoscopic stereotactic technique as a minimally invasive neurosurgical method can also be used for treatment of brain abscess.

Keywords: Brain abscess; endoscopy; stereotaxy.

Introduction

There is still a controversy whether brain abscesses should be treated surgically or conservatively with antibiotics. Different operative techniques for brain abscesses have been established. Abscess puncturing and evacuation through a simple burr-hole was first reported by McEwen and improved by Dandy[8,21]. This procedure was complemented by the use of drain of different consistency for instance rubber catheters[17]. Other operative regimen favor the open microsurgical intervention with abscess evacuation and excision of the abscess membrane, or primary total extirpation within the borders of normal brain tissue, which could include a lobectomy[3,18,19]. Stereotactic aspiration techniques are another approach to brain abscess surgery. Stapleton recently described the stereotactic technique and reported about results[26].

In 1922 King was the first who performed evacuation of a brain abscess under endoscopic control[24]. We have renewed this endoscopic operative approach and combined it with a CT-stereotactic technique for treatment of subacute and chronic membraneous abscesses.

The endoscopic stereotactic evacuation of brain abscesses has three main objectives.

1. Reduction of acute ICP elevation.
2. Sampling of infectious material for microbiological examination.
3. Cure by means of serial puncturing and aspiration.

Methods

1. Instrumentation

For endoscopic brain abscess evacuation we use flexible fiberscopes together with the rigid Marburg Endoscopy Fixation and Guiding System (Martin Co. Tuttlingen). This combination has the advantage to give the endoscope the necessary rigidity for penetration and does not impair its flexibility, which is required for successful abscess evacuation. For stereotactic guidance the CT compatible Mundinger/Birg frame is used. Supplementary instruments for incision of the abscess membrane are microforceps, microscissors and microcatheters with diameters of less than 1.2 mm, which can be inserted through the endoscope's working channel. Haemostasis, if required, is done by ultrathin RF probes (diameter 1.2 mm) or bare laser fibers.

2. Operative Technique

The abscess puncture could be done by free-hand endoscopic or by endoscopic stereotactic technique. We prefer the stereotactic guidance to be absolutely sure of hitting the target, especially in deep-seated processes. CT stereotactic calculation offers data of the target volume and the operative approach, for instance burr-hole co-ordinates, and the way of entry. After a small burr-hole trepanation the flexible endoscope is guided under stereotactic conditions with the help of the Marburg Endoscopy Fixation and Guiding System to the abscess membrane. The membrane, which might be thick and elastic, is perforated or incised with small scissors under endoscopic control. Small haemorrhages, which may occur from the membrane's surface, are controlled by the use of radiofrequency (RF) or Laser-beam. After opening of the membrane a small silicon catheter is inserted through the endoscope's working channel and the contents is aspirated under permanent rinsing with saline solution. At the beginning of this procedure the intra-operative picture quality is reduced by pus and cell detritus. In the further course after extensive rinsing and suction the view will improve and the extent of abscess evacuation can be judged. In this manner it is possible, to evacuate the whole abscess. It is not necessary to place a drain. The decision whether the abscess membrane should be excised in "a second look" operation, depends on the further clinical course of the patient especially taking into account the risk of late epileptic seizures.

Results

From August 1989 to August 1993 a total number of seven patients with brain abscess were treated in the Department of Neurosurgery, Philipps University Marburg by the endoscopic stereotactic technique. There were five males and two females. The age was between 20 and 58 years. In six patients localization was supratentorial, in one patient the abscess was infratentorial intracerebellar (Fig. 1a and b). Microbiological investigations revealed Staphylococcus in four cases, Streptococcus and Bacteroides in one case. In one case it was not possible to establish microbiological diagnosis. The follow up period was between 3 and 48 months. The outcome was measured by the Karnofsky Performance Scale. Six patients showed an improvement of 30% and more compared to the Karnofsky Score determined on admission. One patient suffered postoperatively from seizures. One patient died 11 months after abscess evacuation from sepsis

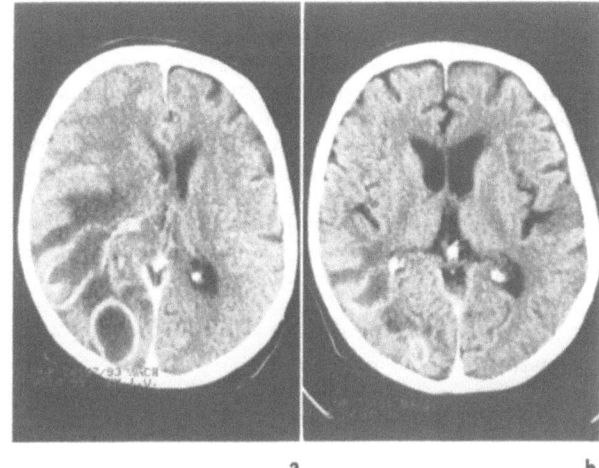

Fig. 2. CCT: multiple right hemispheric brain abscesses (a) before and (b) three weeks after endoscopic stereotactic evacuation and consecutive antiobiotic treatment

caused by bacterial endocarditis. Postoperative CT or MRI control examination showed in all patients a reduction of the abscess diameter (Fig. 2a and b). In 4 cases we performed a second operation after CT had shown a residual abscess cavity with space occupying effect.

Case Report

A 44 year old woman had been hurt by her husband with a bolt shot. She had been admitted to our department in a sleepy state with severe aphasia and slight hemiparesis. The wound and the opened dura mater had been treated initially. Bone however had

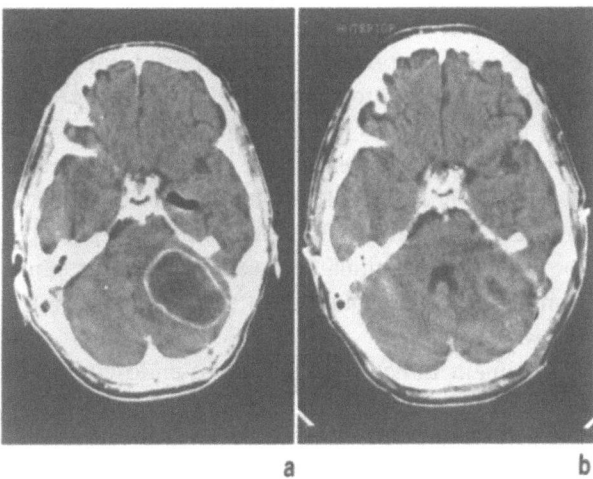

Fig. 1. CCT: (a) left cerebellar brain abscess after chronic suppurative otitis media; (b) after endoscopic stereotactic intervention the abscess size is markedly reduced

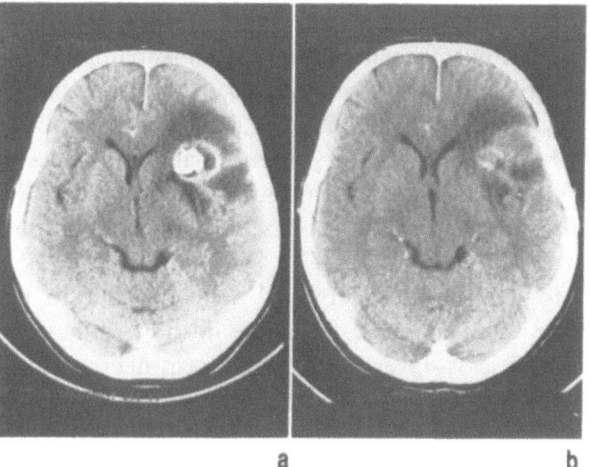

Fig. 3. (a) CCT brain abscess after "bolt-shot" injury with the depressed bone fragment at the bottom of the abscess; (b) postoperative control examination shows the residual abscess

been left in situ deep in the white matter of the frontal lobe. Two months after the injury the patient developed fever and neurological symptoms. CT showed a huge frontal brain abscess with the bony fragment at its bottom (Fig. 3a). CT guided stereotactic endoscopic evacuation of the abscess was performed. First the abscess was punctured. The endoscope was guided stereotactically to the bone fragment. The depressed bone and the attached hair were removed with a special grasping device. Postoperative CT examination shows the remaining abscess membrane (Fig. 3b). Two weeks later the neurological condition was improved and the patient could be transferred to a rehabilitation unit.

Discussion

Etiology and Diagnosis

Brain abscesses can result from a variety of lesions. Chronic suppurative otitis media is the typical source for an "infection per continuitatem"[33]. Today, early diagnosis and consistent treatment of the primary infection lowers the number of brain abscesses caused by otitis. The other important cause in the development of brain abscesses is haematogenous metastatic spread from infectious lung disease[28], cardiac septal defects or bacterial endocarditis[18]. Brain abscesses after open skull injuries[9] and iatrogenic causes after brain surgery are less frequent[25].

With the introduction of CT and MRI diagnosis and localization of brain abscesses has become more easy[23,25], although CT diagnosis may fail in some cases[5,16]. Two cases in our series were admitted to our hospital with the diagnosis of a malignant glioma based on CT examination which was a hazardous misdiagnosis.

Therapeutic Considerations

Despite improvement of diagnostic possibilities, treatment of brain abscesses remains a subject of controversy. To cite F.C. Grant: "Of all the lesions involving the brain, none requires judgement and experience in its successful management more than an abscess"[10]. This statement of Grant has maintained its importance till today.

Horsley[13] and MacEwen[21] were the first to describe their operative technique in the treatment of brain abscess. Over the years different neurosurgical operative strategies have been proposed.

Osteoplastic or osteoclastic trepanation with the "open" evacuation of the pus and the resection of the abscess membrane is the conventional method in the surgery of brain abscess[1,18,19]. Some authors propose to extend the resection into healthy brain tissue[7]. Oth-

ers stress the importance of removing the abscess contents but leaving the abscess capsule intact and in situ in order to reduce the rate of postoperative epilepsy[22]. However the open intervention bears the risk of major damage to brain tissue on the operative route to and around the lesion[26]. Abscesses in deep-seated localizations should not be operated on using the open approach. The mortality of primary exstirpation depends on the state of the abscess. In the chronic state it is around 12 %, in the subacute or acute stage it is estimated to be between 50 and 60 %[31].

A simple less invasive but effective operative method is the aspiration of the abscess contents through a burr-hole[8,27]. The indications for this puncture aspiration method are mainly encapsulated abscesses especially in deep-seated localization (thalamus, brain stem) and in elder patients or patients in poor physical condition[15,30].

Botterell and Dandy[4,8] reported a low mortality rate using this method. However, there are some marked disadvantages associated with free-hand aspiration. The first is the problem of reaching the abscess in a sure-hitting manner. The puncturing probe may fail to perforate the abscess membrane and can be misguided. Some authors describe serious complications using the free-hand puncture-aspiration method[7,19]. The mortality rate of patients operated on with this technique remains still at approximately 50 %[31].

The insertion of drain into the abscess-cavity was first described by Horsley and McEwen[13,21] and improved over the following years with different kinds of material such as gauzes, sponges or rubber-tubes[10].

CT guided stereotactic evacuation of brain abscesses seems to provide some important advantages compared to the procedures described above. Up to now there are many studies about successful stereotactic brain abscess treatment[6,14,20,32]. In his recently published report Stapleton highlighted its effectiveness. CT-localization of the lesion and the passage of the puncturing cannula are accurate, cortical damage is minimized, the incidence of postoperative epilepsy is low[26].

The main indications for neuro-endoscopic interventions are lesions localized in the so-called performed intracranial spaces or cystic brain lesions[11,12]. The brain abscess with its cavity and surrounding membrane is ideally suited to the use of the neuro-endoscopic operative technique. We have combined this with the stereotactic aspiration technique recently described by Stapleton. Postoperative antibiotic cover according to antibiogram is obvious.

Our results are encouraging and emphasize endoscopic stereotactic abscess evacuation as an alternative operative method to conventional techniques.

The main advantages are:

1. Minimized brain damage using a minimal invasive approach.
2. Reliability and safety by CT-stereotactic calculation.
3. Evacuation under direct visual control.

References

1. Arseni C, Ciurea AV, Cirue M (1983) Cerebral abscesses of unknown origin. General report. Zentrbl Neurochir 44: 39–43
2. Ballantine HT, Shealy CN (1959) The role of radical surgery in the treatment of abscess of the brain. Surg Gynec Obstet 109: 370–374
3. Bidzinski J, Koszewski W (1990) The value of different methods of treatment of brain abscess in the CT era. Acta Neurochir (Wien) 105: 117–120
4. Botterell EH, Drake CG (1952) Localized encephalitis, brain abscess and subdural empyema. J Neurosurg 9: 348–352
5. Britt RH, Enzmann DR (1983) Clinical stages of human brain abscesses on serial CT scans after contrast infusion. J Neurosurg 55: 590–603
6. Broggi G, Franzine A, Peluchetti D, Servello D (1985) Treatment of deep brain abscesses by stereotactic implantation of an intracavity device for evacuation and local application of antibiotics. Acta Neurochir (Wien) 76: 94–98
7. Choudhury AR, Taylor JC, Whitackes P (1977) Primary excision of brain abscess. BMJ: 1119–1121
8. Dandy WE (1926) Treatment of chronic abscess of the brain by tapping: preliminary note. JAMA 87: 1477–1478
9. Danzinger, J, Allen KLÖ, Bloch S (1980) An analysis of 113 intracranial infections. Neuroradiology 19: 31–34
10. Grant FC, Groff RA (1938) The surgical treatment of brain abscess. Penn Med J 41: 597
11. Hellwig D, Bauer BL, List-Hellwig E, Mennel HD (1991) Stereotactic endoscopic procedures of the cranial midline. Acta Neurochir (Wien) [Suppl] 53: 23–32
12. Hellwig D, Bauer BL (1992) Minimally invasive neurosurgery by means of ultrathin endoscopes. Acta Neurochir (Wien) [Suppl] 54: 63–68
13. Horsley VL (1988) Case of a cerebral abscess successfully treated by operation: surgical history of the case. BMJ: 636
14. Itakura T, Yokote H, Ozaki F, Itatani K, Hayashi S, et al (1987) Stereotactic operation for brain abscess. Surg Neurol 28: 196–200
15. Jooma OV, Pennybacker JB, Tutton, GK (1951) Brain abscess: aspiration, drainage or excision? J Neurol Neurosurg Psychiatry 14: 308–313
16. Kazner E, Steinhoff H, Wende S, et al (1979) Ring-shaped lesions in the CT-scan. Differential diagnostic considerations. In: Advances in neurosurgery, Vol 6. Springer, Berlin Heidelberg New York, pp 80–85
17. King JEJ (1954) Brain abscess: evolution of the methods of treatment. Ann Surg 139: 587–594
18. Krayenbühl HA (1967) Abscess of the brain. Clin Neurosurg 14: 308–313
19. LeBeau J, Creissard P, Harispe L, Redondo A (1973) Surgical treatment of brain abscess and subdural empyema. J Neurosurg 38: 198–203
20. Lunsford LD (1987) Stereotactic drainage of brain abscess. Neurol Res 9: 25–44
21. McEwen W (1893) Pyogenic diseases of the brain and the spinal cord. Maclehose, Glasgow.
22. Maurice-Williams RS (1987) Experience with "open evacuation of pus" in the treatment of intracranial abscess. Br J Neurosurg 1: 343–351
23. Miller ES, Dias PS, Uttley D (1988) CT-scanning in the management of intracranial abscess: a review of 100 cases. Br J Neurosurg 2: 439–446
24. Nicola N, Sprick C (1987) Der Hirnabszeß. In: Schirmer M (ed) Perimed, Erlangen, pp 90
25. Rosenblum ML, Hoff JT, Norman D, et al (1978) Decreased mortality from brain abscess since advent of computerised tomography. J Neurosurg 49: 658–668
26. Stapleton SR, Bell BA, Uttley D (1993) Stereotactic aspiration of brain abscesses: is this the treatment of choice? Acta Neurochir (Wien) 121: 15–19
27. Stroobandt G, Zech F, Thavoy C (1987) Treatment by aspiration of brain abscesses. Acta Neurochir (Wien) 85: 138–147
28. Tiyaworabun S (1981) 34 year therapeutic experience with brain abscesses: timing problems. In: Advances in neurosurgery, Vol 9. Brain abscess and meningitis. Springer, Berlin Heidelberg New York, pp 48–56
29. Van Alphen HAM, Dreissen JJR (1976) Brain abscess and subdural empyema. Factors influencing mortality and results of various surgical techniques. J Neurol Neurosurg Psychiatry 39: 481–490
30. Vanglider JV, Allen WE, Lesser RA (1974) Pontine abscess: survival following surgical drainage. Case report. J Neurosurg 40: 386
31. Wallenfang T, Reulen HJ, Schürmann K (1981) Therapy of brain abscess. In: Advances in neurosurgery, Vol 9. Brain abscess and meningitis. Springer, Berlin Heidelberg New York, pp 41–45
32. Wise BL, Gleason CA (1979) CT-directed stereotactic surgery in the management of brain abscess. Ann Neurol 6: 457
33. Yang SY (1981) Brain abscess: a review of 400 cases. J Neurosurg 55: 794–798

Correspondence: Dieter Hellwig, M.D., Department of Neurosurgery, Philipps University Marburg, Baldingerstrasse, D-35033 Marburg Federal Republic of Germany.

Acta Neurochir (1994) [Suppl] 61: 106–107

Epiduroscopy and Spinaloscopy: Endoscopic Studies of Lumbar Spinal Spaces

R.G. Blomberg

Department of Anaesthesia, Central Hospital, Norrköping, Sweden

Summary

Previous studies of lumbar spinal spaces using endoscopy—epiduroscopy and spinaloscopy—are recapitulated in this presentation with emphasis on results and their clinical importance for anaesthetic practice.

Keywords: Spinal canal; epidural space; epiduroscopy; spinaloscopy.

Introduction

In 1963, the Dutch neurosurgeon Luyendijk presented radiological evidence of a dorsal fold of the lumbar dura mater and postulated that the dura mater was attached to the dorsal aspect of the epidural space[6]. His results were however questioned in the scientific literature. In an attempt to examine these structures endoscopy of the lumbar epidural space, epiduroscopy, was developed. The method was applied also to the subdural and subarachnoid spaces and then called spinaloscopy. Cadaver studies as well as examinations of patients were performed and published according to the references listed below. The following presentation is a summary of previously published studies.

Methods and Material

For endoscopy a thin, rigid arthroscope, the Olympus Selfoscope, was used, mainly the one with an external diameter of 2.2 mm. An epidural needle was introduced into a lumbar interspace, usually L3–4, and the endoscope into the next interspace cephalad. By injecting small increments of air, the epidural space was expanded. When favourable circumstances were at hand, it was possible to view up to one lumbar interspace in both caudal and cranial directions[1]. The method was used for examination of the lumbar epidural space in autopsy subjects[2] and eventually in living man[3]. Also the lumbar subdural space was studied in autopsy subjects[4]. Finally, techniques for lumbar epidural puncture and catheter introduction were studied in autopsy subjects[5]. The lower thoracic and lumbar subarachnoid space has also been examined in a pilot study and preliminary, unpublished findings are discussed.

Results

The *lumbar epidural space* was examined in 48 autopsy subjects[2]. A dorsomedian connective tissue band consisting of strands of connective tissue with a varying appearance and density was found in each subject. The band caused a fixation of the dura mater to the dorsal aspect of the lumbar space, i.e. the periosteal layer covering the laminae and the flaval ligaments. Thereby, the mobility of the dura mater was restricted. A dorsal fold of the dura mater was also produced. When an epidural catheter was introduced by the midline approach, the tip of the catheter caused tenting of the dura before the catheter bent off, often deviating laterally or caudally.

The lack of circulation as well as the very low or equalized CSF-pressure in autopsy subjects made it necessary to study the structures in living man[3]. The lumbar epidural space was examined in 10 patients prior to a partial laminectomy for a herniated lumbar disc. 8 successful examinations were performed. Partly in contrast to the findings in cadavers, the epidural space in vivo was seen to be mainly a potential space. Influences of a normal circulation and of the pressures of CSF, thorax and abdomen were proposed to account for these differences. The space was easily established by injection of air but returned to its original state as the air disappeared. The dura mater was found to be attached to the dorsal aspect of the epidural space. A dorsal fold of the dura was also seen, however not so prominent as that found in autopsy subjects.

The *lumbar subdural space* was examined in spinaloscopy of 15 autopsy subjects[4]. The space was exposed by injecting saline through the sheath as the endoscope was slowly retracted from the subarachnoid space. In this procedure, the arachnoid membrane easily separated from the dura mater and the otherwise closed but potential space was established. The subdural space was thus found to be a potential space that could easily be established as fluid was injected. It was capable of containing the bevel of an epidural needle. An epidural catheter could also be introduced into the space.

The *techniques of epidural puncture* with the midline and paramedian approaches were compared by epiduroscopy in 14 autopsy subjects[5]. The angle of the needle to the dura mater was near perpendicular with the midline approach. In the paramedian approach the angle was found to be 120°–135°. Each needle was advanced to contact with the dura mater and the risk of dural perforation with further advancement judged. An epidural catheter was also introduced through each needle, one at a time, and the course of the catheter described. There was an imminent risk of dural perforation for all 14 midline needles with further advancement. The risk was imminent for only two and possible for one needle with the paramedian approach. The catheter introduced with a midline approach caused tenting of the dura in all 14 instances and only 4 of these catheters took a straight course cephalad. The catheter passed through a needle by the paramedian approach did not cause tenting in any case and went on in a straight course cephalad in all 14 instances. The common factor in the behaviour of needle and catheter was seen to be the difference in the angle to the dura mater in combination with the influence of the restricted mobility of the dura mater.

Preliminary and unpublished data of *spinaloscopy* of the lower thoracic and the lumbar subarachnoid space were presented. Special attention was paid to connective tissue attachments causing restriction of nerve root mobility and to any possible association with complications of subarachnoid puncture.

Discussion

The studies presented here have demonstrated the effect of a dorsal fixation of the dura mater in the lumbar epidural space called the dorsomedian connective tissue band. In restricting the mobility of the dura the band was shown to contribute to the risk of dural perforation in epidural puncture when the midline approach was used. Also, the introduction of an epidural catheter with the midline approach was influenced by the restricted mobility of the dura. Tenting of the dura by the tip of the catheter was often caused. The catheter course was then frequently deviated laterally or caudally.

References

1. Blomberg R (1985) A method for epiduroscopy and spinaloscopy. Presentation of preliminary results. Acta Anaesthesiol Scand 29: 113–116
2. Blomberg R (1986) The dorsomedian connective tissue band in the lumbar epidural space of humans. Anesth Analg 65: 747–752
3. Blomberg RG, Olsson SS (1989) The lumbar epidural space in patients examined with epiduroscopy. Anesth Analg 68: 157–160
4. Blomberg RG (1987) The lumbar subdural extraarachnoid space of humans. Anesth Analg 66: 177–180
5. Blomberg RG (1988) Technical advantages of the paramedian approach for lumbar epidural puncture and catheter introduction. Anaesthesia 43: 837–843
6. Luyendijk W (1963) Canalography: röntgenological examination of the peridural space in the lumbosacral part of the vertebral canal. J Belge Radiol 46: 236–254

Correspondence: Rune G. Blomberg, Ph.D., M.D., Department of Anaesthesia, Central Hospital, S-601 82 Norrköping, Sweden.

Acta Neurochir (1994) [Suppl] 61: 108–114

Operative Spinal Endoscopy: Stereotopography and Surgical Possibilities

V.B. Karakhan, B.A. Filimonov, Y.A. Grigoryan, and **V.B. Mitropolsky**

Department of Neurology and Neurosurgery, Moscow Medical Stomatological Institute, Moscow, Russia

Summary

The polyprojective microstereotopography of spinal canal structures at the cerebello-spinal, cervical, thoracic, lumbosacral and cauda equina levels on 20 fresh cadavers is presented using flexiscopes 3.7–3.9 mm diameter. This is possible due to the space between spinal cord–vertebral canal which is about 10 mm at all levels. This also allows one to insert the endoscopic tube by posterior or interradicular approach. The subdural and subarachnoid endoscopic examinations have been performed through small foraminotomic openings with resection of the base of the spinous process. The anterior and posterior roots, the spinal cord, dural root sleeves, cerebellar tonsils, orifice of the IV ventricle, vertebral artery and its lower branches can be visualised.

On the stereotopographic basis the first operations in patients with severe spinal cord injury (detection of multilevel cord compression, removal of massive subarachnoid bleeding), syringomyelia and haemorrhage into the IV ventricle (clot removal by the ascending cervical route) were undertaken. More than 10 real and probable indications for operative spinal endofiberoscopy are discussed.

Keywords: Spinal canal; endoscopy; surgery.

Introduction

The examination of lumbar part of the spinal canal using the need endoscope is well-known[1-4,7,9-12]. New surgical possibilities can be developed with the insertion of endofiberscopic tube through the microsurgical opening into spinal canal. But this canal already contains the natural tube—a spinal cord. It is the challenge to the nature, isn't it?

Stereotopometric data (see Table 1) show that average anterior-posterior (A-P)—lateral (lat) dimensions of the spinal cord and vertebral canal at all levels are distinguished in ten mm or more in adults. This justifies a multilevel examination of spinal canal by endofiberscopes of 3–4 mm tube diameter.

Material and Methods

20 fresh adult cadavers have been investigated endoscopically and endoscopic procedures were performed in 5 patients with the severe spinal cord injury—2 (suspicion on multilevel cord compression, massive subarachnoid sac bleeding), syringomyelia-1, spinal arachnoidal cyst-1, and retained clots in IVth ventricle-1.

Flexiscopes with bending tip were used: ENF-P ("Olympus") with 3.7 mm tube diameter and PF-893 ("Aesculap") with 3.9 mm one including 2 channels.

Investigations were performed through the microsurgical opening with removal the base of spinous process, dural and arachnoidal incisions and withdrawal of some CSF.

Traces of endoscopic examination (see Fig. 1a) were based on key-directing structures (neural roots, dural sleeves etc.) ensuring the precise flexiscope navigation. The level of observation was detected by measurement of the length of endoscope insertion.

Results

Endoscopic traces stereotopography. Ascendent and descendent accesses were used. At the cervical and thoracic levels 3 approaches are outlined: posterior, interradicular and preradicular (Fig. 1a). Lateral diameter at the cervical level is predominant. Therefore one can pass flexiscope between roots without difficulties.

Cervical endofiberscopy. By lateral (interradicular) approach endoscopic images (Fig. 1 b,c) allow to detect posterior, and anterior spinal roots, their entry zones and branches, dural sleeves, dentate ligaments,

Table 1. *Average Anterior-Posterior (A-P) and Lateral (Lat) Diameters of the Spinal Cord and Vertebral Canal in Adults*

| Segmental level | Diameters (mm) | | | |
| | Cord[a] | | Canal[b] | |
	A-P	Lat	A-P	Lat
Cervical (C 5–6)	7.2	13.2	14.7	24.5
Thoracic (Th 6–7)	6.5	8.0	16.8	17.2
Lumbar (L 5)	8.0	9.6	17.4	23.4

[a] Elliot HC (1945); [b] Aeby C (1879).

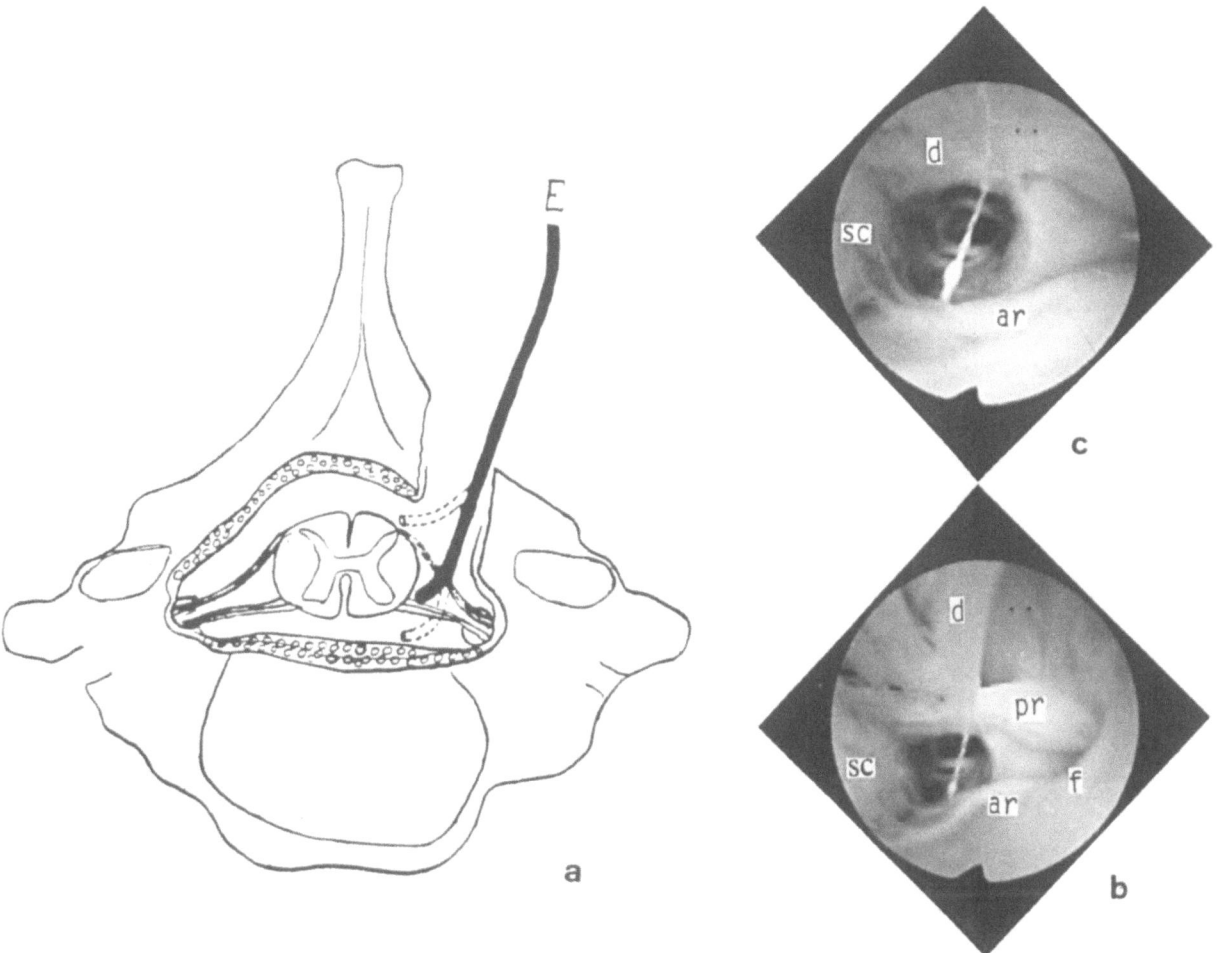

Fig. 1. The route of flexiscope *(E)* insertion at the cervical level (a) and endoscopic images (b, c) of the right lateral aspect of the spinal cord *(sc)*. The serial subarachnoid view in cranial direction (b → c). *d* denticulate ligament, *f* dural root sleeve, *ar, pr* anterior and posterior roots. In center—arachnoid strand

spinal cord and inner dural surface. Posterior approach ensures recognition of the spinal cord relief, its vascular network posterior roots, and its entry zones, middle arachnoid septum. Trigon-like contours of vertebral arc helps to stereoorientation. Upward directions of this access lead to craniospinal joint area.

Cranio-spinal joint endoscopy (Fig. 2). Assessment of the lateral aspect of craniospinal zone allows to detect the trace of the vertebral artery, beginning of the posterior spinal and posterior inferior cerebellar artery, lower group of cranial nerves. Examination of posterior aspect of craniospinal zone shows the posterior arc of the foramen magnum, cerebellar tonsils and approach to IV ventricle between the tonsils and the medulla oblongata. Endoscopic orientation within cerebello-spinal region ensures the detection of clots protruded from IV ventricle between cerebellar ton-

Fig. 2 (see p. 110). Endofiberoscopic stereotopography of the craniospinal joint zone (ascending subarachnoid postero-lateral approach). (a–c) Endoscopic images and sketch of the right postero-lateral aspect of the cerebello-spinal junction. *M* edge of foramen magnum, *t* cerebellar tonsil, *sc* spinal cord, *a* arachnoid strands. Arrows on the sketch show the traces of endoscope passing toward middle (d, e) or lateral (f, g) parts of cerebello-spinal junction. (d) Both tonsils, medulla oblongata and orifice of the IV ventricle between them are determined; (e) more upward and lateral view: edge of foramen magnum and supracerebellar space in addition are seen; (f) dural penetration of the vertebral artery (VA) in contact with medulla oblongata and the posterior inferior cerebellar artery (short arrow). On the posterior frontal plane the direction of endoscope *(E)* insertion is outlined

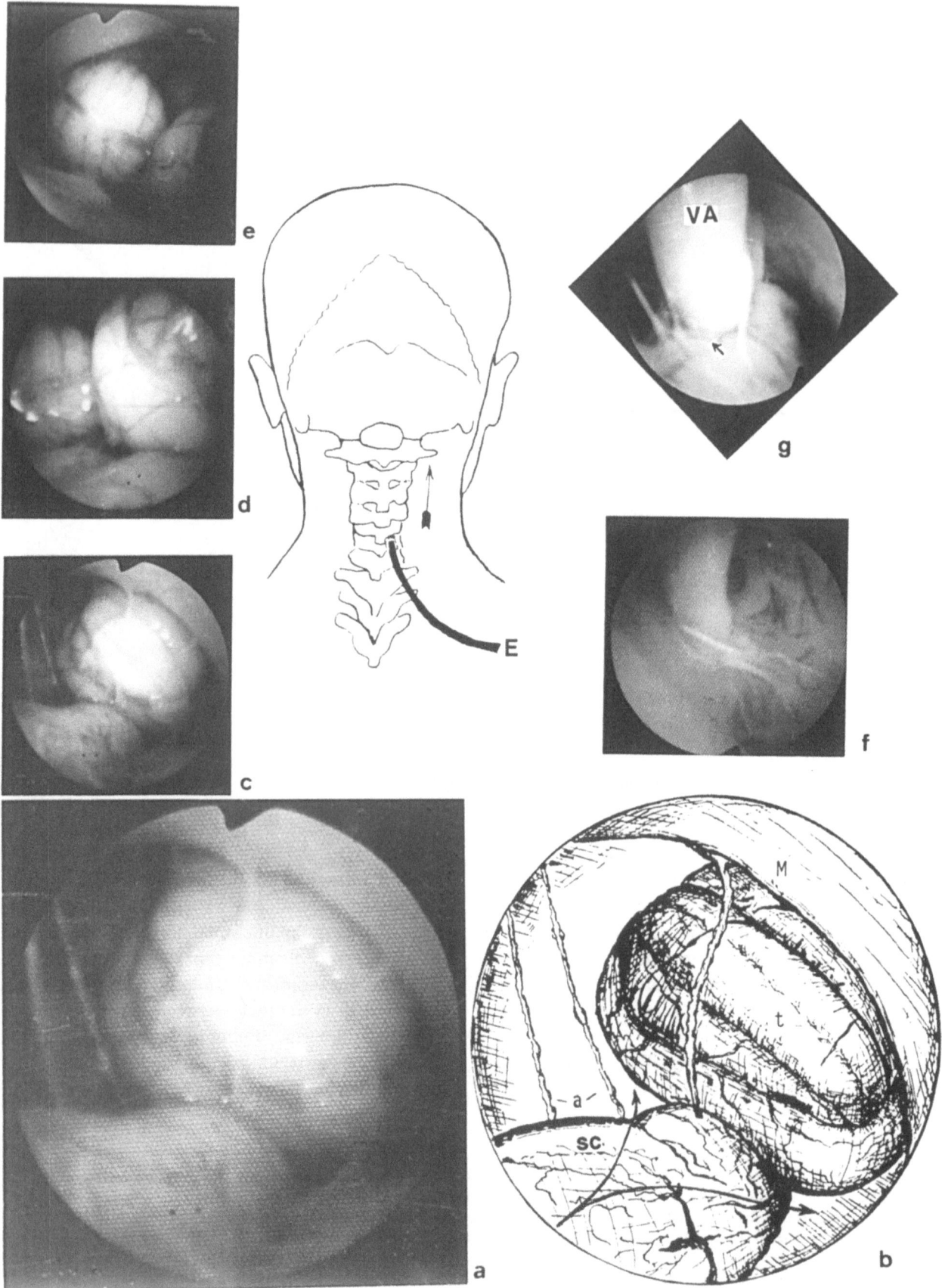

Fig. 2. Legend see p. 109

sils and removal of blood. For penetration into IV ventricle endoscopic tip is inserted between uvula, obex and loops of the posterior inferior cerebellar artery. One can see the floor relief, middle fissure, leading to aqueduct orifice and aqueductal canal. On the other side the descendent approach via anterior horn with perforation of premammillar area of the IIIrd ventricle[8] allows one to penetrate into middle pontine cistern and then to anterior arc of the foramen magnum, hypoglossal nerve and upper cervical anterior roots.

Thoracic endofiberscopy (Fig. 3a–d). Lateral aspect of visualization by descendent or ascendent approach shows a typical roof-shaped roots configuration. Assessment of regional stereotopography including also dentate ligaments, dural sleeves and spinal cord ensures a strict displacement of the roots by endoscopic dissector.

Posterior approach allows to detect the oblique radicular entry to the cord, posterior arachnoid septum, vascular network on the surface of the cord and a gap between it and dural surface. In these conditions endoscopic tube might be easily inserted in 100 or even 200 mm beyond the opening site. Spinal block is recognized by disappearance of the subdural gap in polyprojective examination. Block level is detected by length of the tube insertion.

Lumbar and cauda equina level (Fig. 3e–h). Through the small opening one can observe lumbosacral part of spinal canal down to terminal segment of the sac, presented in form of sunbeams. Assessment of contours of lumbar canal ensures a stereoorientation (anterior, posterior walls of canal) among the cauda roots. Dural sleeves and filum terminale also are visualized. Root configuration depends on the level of examination (Fig. 3f–h). Arachnoidal cyst is detected by prolapsed arachnoid wall and adjacent root displacement.

Spinal subarachnoid sac bleeding endoscopically is presented as a dark area. After endoscopic blood aspiration stereotopography becomes clear.

Intraspinal examination (Fig. 4). In syrinogomyelia insertion of endoscopic tube 3.7 mm diameter into a wide cavity allows to investigate it, remove a fluid, dissect adhesions avoiding cord injury with good clinical outcome. This case shows the possibility of damageless insertion of the endoscopic tube into unseptated wide syrinx cavity.

Discussion

Stereotopographic data and preliminary clinical results of flexiscopes usage ensuring good visualization and actions (mobilization, dissection, perforation, aspiration) justify the real and probable indications for operative spinal endoscopy (Table 2).

1. We suppose that spinal access to IVth ventricle is more simple than transoccipital in urgent cases and may be useful for removal of remnant clots from IVth ventricle in cases of wide-spread ventricular haemorrhages.
2. Precision of multilevel compression and withdrawal of the sac blood during operation in severe cord injury may be the useful application of spinal endofiberscopy.
3. Although the advantages of endoscopy for treatment of hydromyelia already have been reported[5,6], coaxial insertion of thick and ultrathin

Table 2. *Real and Probable Indications for Operative Spinal Endoscopy*

Real
1. Clots in the IV ventricle
2. Multilevel traumatic cord compression and large subarachnoid haemorrhage
3. Syringomyelia
4. Spinal arachnoid cysts and adhesions
5. Meningoradiculocele
6. Transcanal removal of disc prolapses

Probable
7. Selective motor roots or DREZ incisions (for spasticity or pain)
8. Transdiscal (transvertebral) lumboperitoneal shunting
9. Ventriculospinal shunting (through the IV ventricle or middle pontine cistern)
10. Correction of acute tonsillar dislocation
11. Tethered cord (filum terminale incision)
12. Spinal radiculo-vascular compression

Fig. 3 (see p. 112). Endofiberscopic stereotopography of thoracic (a–d) and lumbosacral–cauda equina (e–h) levels (a, b, d–h—posterior approach, (c) interradicular approach). (a) view of postero-lateral surface of the spinal cord (*sc*) and roots at the upper thoracic level on the left; (b) delineation of the posterior arachnoid septum and superficial veins (*D* dural surface); (c) roof-shaped configuration of the anterior and posterior (*pr*) roots at the upper thoracic level (*d* denticulate ligament, *f* dural sleeve); (d, e) oblique position of the spinal roots at lower thoracic and upper lumbar levels (*sc*—spinal cord); (f–h) cauda equina visualization at the upper, middle and terminal sac levels. The plane in profile shows a general topography of examination

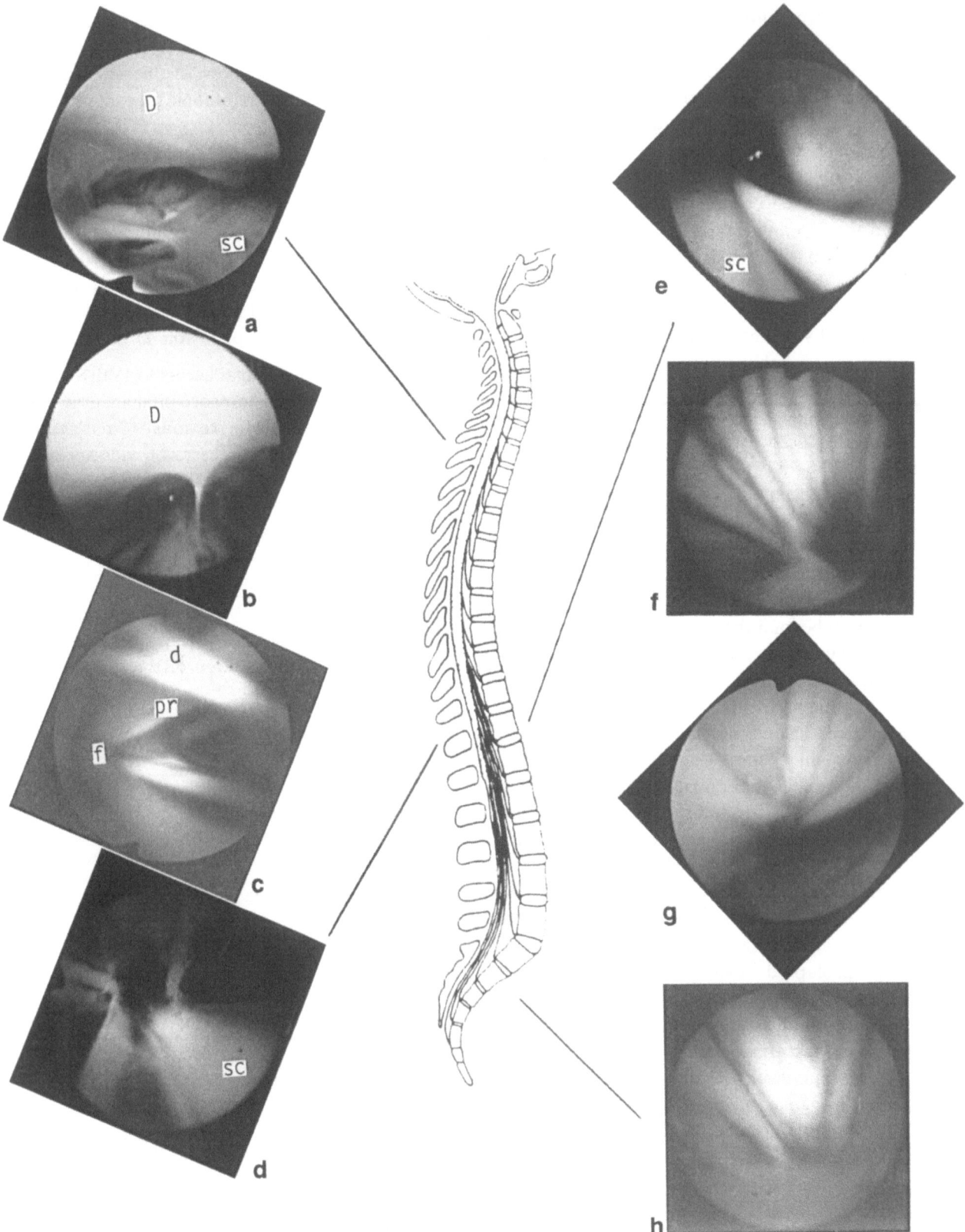

Fig. 3. Legend see p. 111

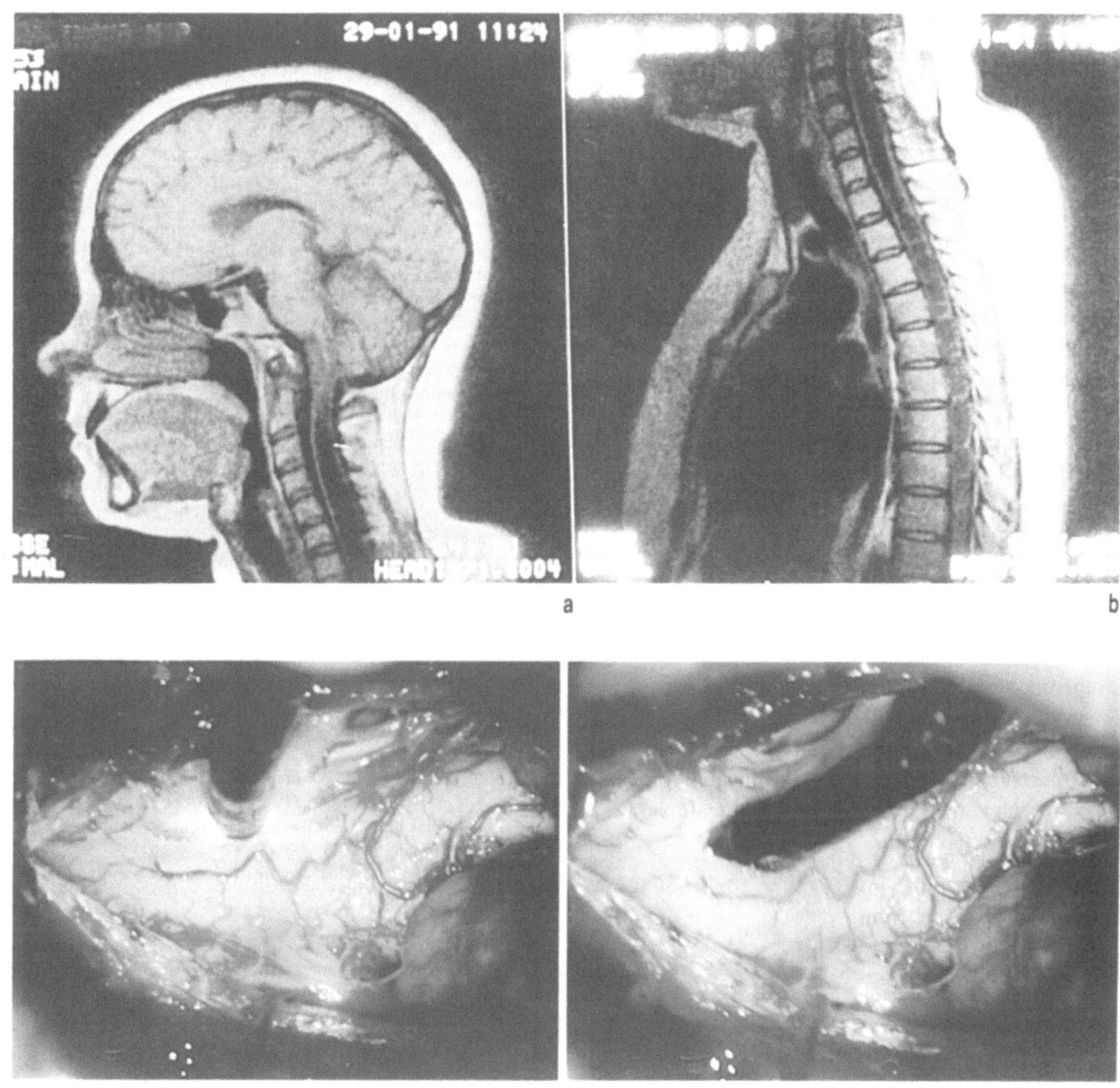

Fig. 4. MRI (a, b) and microsurgical examination (c, d) in the case of large unseptated syringomyelia. (c) Insertion of the flexiscope 3.7 mm diameter into syrinx cavity (bright halo around the endoscopic tip is visualized); (d) intracavital advancement of the endoscopic tube

endoscopes provide the significant decrease of sizes of surgical access.

4. The use of spinal endofiberscopy in cyst pathology also may diminish the operative approach with wide incision of cystic walls.

5. Intraoperative endoscopic examination ensures precise surgical strategy in meningoradiculocele.

6. Usefulness of endoscopy for transcanal removal of prolapsed discs is in control and clearance of interbody cavity after removal of ruptured disc and in withdrawal of displaced sequestrated prolapses.

7. Multilevel visualization of anterior or posterior root/lets ensures the selective incision of them endoscopically.

8. Development of direct endoscopic way and devices for communication between spinal canal and posterior part of peritoneal cavity avoiding the omentum may solve the problem of malfunctioning of lumboperitoneal shunts.

9. Anterior or posterior shunting avoiding blocked infracerebellar or interpeduncular cisterns using ultra thin flexiscopes can improve draining techniques (spare procedures directing the CSF flow in liquor system).

Stereotopographic data show that endofiberscopy indeed may unite intracranial and spinal spaces.

10. Visualization of descendent cerebellar tonsils by spinal endoscopy and development of method of selective endoscopic correction of tentorial pressure cone[8] justify the trends of endoscopic insertion of herniated tonsils.

11. Endoscopic identification and selective incision of the filum terminale and arachnoid adhesions may correct the tethered cord syndrome with minimal surgical trauma.

And finally,

12. First of all it's necessary to discover the existance of neuroradicular pathologic cross-contacts by means of polyprojective endoscopic examinations on randomized cadavers. Then it might be recommended the additional spinal subdural flexiscopic assessment in cases with clear delineated low back pain syndrome when epidural surgical examination isn't satisfactory.

These points are the investigating topics of spinal endofiberscopy in the future.

Acknowledgement

The authors are indebted to A. Voronov for sketch drawing.

References

1. Blomberg R (1985) A method for epiduroscopy and spinaloscopy. Presentation of preliminary results. Acta Anaesthesiol Scand 29: 113–116
2. Döhring S, Ooi Y, Schulitz KP, Satoh Y (1984) Myeloskopische Befunde im Bereich der unteren lumbalen Wirbelsäule. Beitr Orthop Traumatol 31: 120–126
3. Forst R, Hausmann B (1983) Nucleoscopy—a new examination technique. Arch Orthop Trauma Surg 101: 219–221
4. Fukushima T, Schramm J (1975) Klinischer Versuch der Endoskopie des Spinalkanals: Kurzmitteilung. Neurochirurgia 18: 199–203
5. Hellwig D, Bauer BL (1992) Minimally invasive neurosurgery by means of ultrathin endoscopes. In: Bauer BL, Hellwig D (ed.). Minimally invasive neurosurgery—MIN. Acta Neurochir (Wien) [Suppl] 54: 63–68
6. Huewel N, Perneczky A, Urban V, Fries G (1992) Neuroendoscopic technique for the operative treatment of septated syringomyelia. Acta Neurochir (Wien) [Suppl] 54: 59–62
7. Karakhan VB, Sokolinsky AV (1986) Spinal endoscopy. Vopr Neirolchir N 3: 48–51
8. Karakhan VB (1992) Endofiberscopic intracranial stereotopography and endofiberscopic Neurosurgery. Acta Neurochir (Wien) [Suppl] 54: 11–25
9. Mayer HM, Brock M, Berlien HP, Weber B (1992) Percutaneous endoscopic laser discectomy (PELD). A new surgical technique for non-sequestrated lumbar discs. Acta Neurochir (Wien) [Suppl] 54: 53–58
10. Neretin VY, Kiryakov VA, Frolov YA, Lobov MA, Kotov SV (1984) Spinal endoscopy in diagnosis of spinal cord lesions. Zhurn Nevropatol Psikhiatr 84: 23–26
11. Ooi Y, Mita F, Satoh Y (1990) Myeloscopic study on lumbar spinal canal stenosis with special reference to intermittent claudication. Spine 15: 544–549
12. Pool JL (1942) Myeloscopy: intraspinal endoscopy. Surgery 2: 169–182

Correspondence: V.B. Karakhan, M.D., Department of Neurology and Neurosurgery, Moscow Medical Stomatological Institute, Delegatskaya str. 20/1, 103473 Moscow, Russia.

Subject Index

U. Ito, A. Baethmann, K.-A. Hossmann, T. Kuroiwa,
A. Marmarou, H.-J. Reulen, K. Takakura (eds.)

Brain Edema IX

1994. 281 figs. XV, 590 pages.
Cloth DM 330,–, öS 2310,–
Reduced price for subscribers to "Acta Neurochirurgica":
Cloth DM 297,–, öS 2079,–
ISBN 3-211-82532-0
(Acta Neurochirurgica / Supplement 60)

This volume is an up-to-date report on progress in the understanding of brain edema, with a spectrum reaching from most recent molecularbiological findings to respective clinical developments. Major topics deal with (a) the blood-parenchymal cell border under normal and pathological conditions causing brain edema, (b) neuronglial interactions and their disturbances in tissue damage, (c) formation, propagation and resolution of brain edema, and finally (d) treatment of vasogenic and cytotoxic brain edema. In the basic science approaches emphasis is given to newly discovered molecules, such as vascular endothelial growth factor, which might control permeability of the blood-brain barrier, e.g. in brain tumors. The complex issue of mediator compounds of secondary brain damage is further developed as to its manyfold involvement, for example in barrier dysfunction, cell swelling, disturbances of the microcirculation, and others. The report further contains comprehensive assessments of edema pathophysiology by advanced technologies, such as in-situ hybridization on the one hand side or NMR-diffusion imaging on the other. Novel forms of treatment acquiring increasing specificity represent a central focus.

Prices are subject to change without notice

Sachsenplatz 4–6, P.O.Box 89, A-1201 Wien · 175 Fifth Avenue, New York, NY 10010, USA
Heidelberger Platz 3, D-14197 Berlin · 3-13, Hongo 3-chome, Bunkyo-ku, Tokyo 113, Japan

A. W. Unterberg, G.-H. Schneider, W. R. Lanksch (eds.)

Monitoring of Cerebral Blood Flow and Metabolism in Intensive Care

1993. 76 figures. VIII, 125 pages.
Cloth DM 140,–, öS 980,–
Reduced price for subscribers to "Acta Neurochirurgica":
Cloth DM 126,–, öS 882,–
ISBN 3-211-82484-7

(Acta Neurochirurgica / Supplement 59)

Until recently the determination of cerebral blood flow was limited to intermittent measurements and particulary difficult to perform in critically ill patients. Monitoring of cerebral blood flow was an unreached goal.

This book gives an update of the methods nowadays available to measure and monitor cerebral blood flow and cerebral oxygenation, especially in critically ill patients. In a first chapter various techniques to measure cerebral blood flow discontinuously are discussed, like the Kety-Schmidt technique, 133-Xenon-CBF and CT-CBF with stable synon. Thereafter various methods to continuously estimate cerebral blood flow are described, like thermodiffusion techniques, monitoring of tissue pO2, dopplersonography, laser-doppler-spectropscopy, near infrared spectroscopy and monitoring of oxygen saturation in the jugular bulb. The last method is highlighted and extensively discussed with its different indications.

Prices are subject to change without notice

Sachsenplatz 4–6, P.O.Box 89, A-1201 Wien · 175 Fifth Avenue, New York, NY 10010, USA
Heidelberger Platz 3, D-14197 Berlin · 3-13, Hongo 3-chome, Bunkyo-ku, Tokyo 113, Japan

Springer-Verlag
and the Environment

WE AT SPRINGER-VERLAG FIRMLY BELIEVE THAT AN international science publisher has a special obligation to the environment, and our corporate policies consistently reflect this conviction.

WE ALSO EXPECT OUR BUSINESS PARTNERS – PRINTERS, paper mills, packaging manufacturers, etc. – to commit themselves to using environmentally friendly materials and production processes.

THE PAPER IN THIS BOOK IS MADE FROM NO-CHLORINE pulp and is acid free, in conformance with international standards for paper permanency.